# THE COLD HARD FACTS ABOUT LDAC

## Leadership Insights and Accessions Guide

Printed in the United States of America.

ISBN: 978-1-59571-454-1
Library of Congress Number: Applied for

Designed and published by:

Word Association Publishers
205 Fifth Avenue
Tarentum, Pennsylvania 15084

www.wordassociation.com
1 800 827 7903

# THE COLD HARD FACTS ABOUT LDAC

LEADERSHIP INSIGHTS AND ACCESSIONS GUIDE

**RODNEY CRUCE**

ROTC INSTRUCTOR OF THE YEAR

# Table of Contents

## Acknowledgments

When I decided to write this book, I had no idea how difficult or how easy it would be to correlate the information in a presentable format that would provide a valuable tool for Army ROTC Cadets attending the Leadership Development and Assessment Course (LDAC). At first, I thought, "How hard can it be?" After all, I had over 20 years of military service writing Operations Orders that provided guidance and information. Afterwards, I realized just how hard it was finally to complete the book. It takes time, forethought and dedication to completing the task. For those who have worked with me or known me first hand, I am an optimist and a realist in the same breath. Once I set my mind to accomplishing a task, come hell or high water, I will complete it and it will succeed. The only part that I did not plan on with this book is the fact it would take me a year to finish the product.

The most important factor I kept in mind while writing this book was to address key information, cold hard facts and leadership insights that Cadets must understand to properly prepare for attending their commissioning course. Over the course of several months the book design and contents have been reorganized several times. The four most significant variables that assisted me in accomplishing the task were time management, experience, influencing individuals and the end user: Cadets!

As I was preparing to retire from the U.S. Army, I spent every moment either working, studying to finish a graduate degree or thinking about what profession I was to pursue. During the final months of my service obligation, my time was the most precious resource required to write the book. Time management, similar to LDAC, was the driving factor. Fortunately, I had gained the skill of time management throughout my military career by assisting other countries, dealing with different personalities and the ability to multi-task. As a member of Special Forces, you are required to multi-task every part of the day to include explaining the task in a foreign language while assisting foreign militaries. I had planned on writing the book prior to retiring from the military but even with all my multitasking experience I did not make the deadline.

Many people have collaborated with me on the research, organization and presentation of information in this book. I am most grateful to LTC Bryce R. Christensen, CSM Martin Celestine, CSM Todd Svenson, SFC Mario Volpe, SFC Jackie South and the remainder of the Cadre at The University of Central Florida (UCF) ROTC Program for the motivation, support and pursuit of perfection. LTC Bryce R. Christensen, whom I consider a great friend as well as a great boss, gave me the latitude and freedom to operate and most of all, the sanity to stay on azimuth in the organization of this book to ensure all pertinent information was addressed.

I am thankful to the many different Non-Commissioned Officers that I worked with during my tenure at the UCF Army ROTC Program for assisting in the implementation and operation of an absolutely outstanding ROTC program that excelled in all aspects and welcomed any challenge. Every one of these NCOs is more than just a colleague to me, as we have all shared great moments and experiences that never interfered with or threatened our work statute. Finally, as a prior Non-Commissioned Officer, I appreciated the honesty and many thought-provoking conversations with me on the best method to train Cadets as future leaders of the Army.

I am indebted to the many Cadets who I was lucky enough to teach, counsel and mentor during my assignment at UCF Army ROTC. You have taught me just as much as I have taught you. As I wrote this book, I referred back to the many different leadership challenges, real world experiences, military education, ROTC classes and labs and most of all, the LDAC evaluation model to provide a complete and relevant product that can be used to prepare Cadets to excel and gain LDAC credit at the commissioning course.

I have been blessed with many wonderful mentors, teachers, peers and operators in my life. I certainly cannot name them all. A few of them from my career with Special Forces do stand out – BG Ed Reeder, LTC William "Bill" Kearney, LTC Anthony Fletcher and every Team Member of SFODA 784, 721 and 797 that I have had the luxury to serve with on a Special Forces Operational Detachment – Alpha (SFODA). They have inspired me to be a leader, a quiet professional, a team player and most of all, to have the honor to serve and be part of the "Family Business."

Since this is my first book, I have learned an abundance of information about the publishing process and realize no book project comes to life without a great publishing team. I would like to express my gratitude to Dr. Tom Costello and Gina Datres at Word Association Publishers. The company led me through the publishing process, was very user friendly while answering all of my questions, and even provided direction when I lacked experience about specific publishing details.

Finally and most importantly, I want to thank my wife, Jaime. She is my inspiration and best friend, who provided me the motivation, great company and conversation that lit the fire in me to write this book. She has given me the balance I need in life and always asks the hard questions that make me think through problems and perform to my full potential. She supports me unconditionally, and for that I am more privileged than I ever thought was possible. Finally, I dedicate this book to the future leaders of the Army. As you begin your journey and career in the military, read this book with the eagerness of learning followed by setting goals for excellence, never take no for an answer and always have the attitude that failure is not an option. These were the ingredients for a successful career and the ones that I live by today. As a former Officer and leader in the Army, I now hand the torch to you through the knowledge in this book as you take the first step of many toward becoming tomorrow's great leader.

# About the Author

Rodney J. Cruce is the CEO/Owner of On Point Saliency (OPS) in Orlando, Florida, where he specializes in Leadership, Managerial Decision-Making, Intelligence Services, International Security and Team Building. He created OPS after retiring from the Army as a Major and since has traveled back and forth to Iraq and Afghanistan assisting the Department of Defense in their overseas operations.

Major (R) Cruce completed over 20 years of successful military service that included 10 years as a Non-Commissioned Officer and 10 years as an Officer. He entered the Army as a Combat Engineer enlisted soldier assigned to the 10th Engineer Battalion, 3rd Infantry Division at Kitzingen, Germany. In 1991, he was transferred to Fort Bragg, NC where he was assigned to the 27th Engineer Battalion, 20th Engineer Brigade until he volunteered for Special Forces Training. After completing the Special Forces Engineer Qualification Course, he was assigned to Special Forces Operational Detachment – Alpha (SFODA) 784, a Military Free Fall Team, in 3rd Battalion, 7th Special Forces Group (7th SFG (A)) from 1993 to the end of 1998. In 1995, he attended Ranger School where he graduated as the Enlisted Honor Graduate. In 1996, he changed his Special Forces specialty from Engineer to Intelligence Sergeant but remained on the same SFODA. During his time at 7th Special Forces Group, he conducted missions in Honduras, Nicaragua, El Salvador, Venezuela, Peru, Colombia and Chili.

In 1998, SFC Cruce was selected to attend Officer Candidate School (OCS) at Fort Benning, Georgia, where he commissioned as a 2LT in Military Intelligence. At OCS, he graduated on the Commandant's List and received the Physical Fitness Award. Following OCS, he attended the Military Intelligence Officer Basic Course where he received the General Potts Leadership Award for physical fitness, academic and leadership excellence. In 2000, 1LT Cruce graduated summa cum Laude from Campbell University with a Bachelor of Science Degree in Computer Science. Afterwards, he returned to 7th SFG (A) as an Intelligence Officer for a brief time before attending the Military Intelligence Captain Career Course. During this time, 1LT Cruce was selected to return to Special Forces and attend the Special Forces Detachment Officer Qualification Course. Upon his completion of the course, CPT Cruce returned to 7th SGF (A) from 2002 to 2006 where he commanded SFODA 721, served as the Assistant Operations Officer (AS3) for the Group Headquarters and commanded SFODA 797. He conducted missions in Afghanistan, Nicaragua, Colombia, Surinam, Venezuela, Curacao, Bolivia and El Salvador. In 2006, MAJ (R) Cruce was assigned to the University of Central Florida ROTC Program where he completed his military service. While at UCF Army ROTC, one of the top programs in the nation, he served as the Assistant Professor of Military Science specifically teaching the MS III Cadets in preparation for LDAC to include completing a Graduate Degree in Industrial Engineering Management from the University of Central Florida. While at UCF Army ROTC, he developed and implemented techniques from his Special Forces Training that produced Cadets who ranked in the top 1% of the nation. Finally, he received The Colonel Leo A. Codd Award in 2007 for The Most Outstanding U.S. Army ROTC Instructor and received the 2008 Cadet Command Instructor of the Year Award.

MAJ (R) Cruce resides in Orlando, Florida with his wife, Jaime, and their highly unstable and ferocious dog, Berkley.

# Preface

In March of 2008, I was at a Brigade Field Training Exercise (FTX) and realized not all Cadets received the proper instruction to be successful at LDAC. LTC Christensen and I began to talk and realized that there was a need for Cadets to receive additional training outside of ROTC if Cadre did not have the previous experience of attending LDAC or were inexperienced. I agreed with this because as a Platoon Tactical Officer (PTO) at LDAC during 2007, there were wide ranges of experience levels that made it apparent that not all Cadets received the same information before attending LDAC.

After the Brigade FTX, my wife and I had a discussion about opening a leadership and team building business focusing on ROTC Cadets and Corporate Executives after retirement from the Army. Over the next several discussions, the idea of writing a book surfaced and made more sense with the economy entering a deep recession. I must say, when she suggested that I write a book, I was a bit mesmerized with the thought because I never thought that I might be an author of a book. Over the next couple of weeks of contemplating and a couple of visits to the 19th hole, the decision had been made. I would write a book that focused on the Cold Hard Facts about LDAC, a guide for Leadership Insights and Accessions to promote success while not attempting to teach the entire MS III Year. As Paul Harvey used to say, "and now, the rest of the story."

When I was a PTO, approximately 20% of Cadets had no idea of how many dimensional excellence marks were required to receive an overall "E" rating. There are many different reasons why these Cadets were uninformed or failed to recall the parameters of the Leadership Development Program (LDP). Some Cadets told me they were never briefed on these requirements, others said they missed that block of instruction, others blamed their instructors for not informing them of the standards, and some just did not understand. Of course there are three sides to every story, yours, mine and the real truth. For the most part, I believe there is a combination of reasons why a significant percentage of Cadets arrived at LDAC without the knowledge of the LDP. First of all, there just is not enough time allotted to teach Cadets every intricate detail about the LDP. Second, some Cadets don't put a lot of effort into learning the process because they are balancing multiple classes and tests during their MS III year, and third, not all Cadre understand the LDP so they may not teach it using the most effective means. Of course there are a thousand other reasons that Cadets may not have received the instruction but in the end, if Cadets are proactive and want to learn, they will seek out the information from a peer, instructor or another Cadre member.

For those Cadets that do not have the awareness or the ability to see they are lacking the necessary information, I hope they are reading this book.

Now this is where this book comes into play. Since Cadets arrive at LDAC with little insight or a game plan on how to succeed or accomplish the milestones required of them at the commissioning course. I saw the need to help those Cadets who failed to remember, failed to understand or were not afforded the opportunity to learn the insights of LDAC before departing their home campus for one reason or another. Believe me, there are a thousand reasons why a Cadet does not remember or did not receive the proper information, knowledge or training before attending LDAC but the truth to the matter is, there is a void and I am filling the gap. Have you ever heard of the age-old adage "knowledge is power"? If not, there is truth to that saying and it lies within this book to include the Cold Hard Facts and Leadership Insights to LDAC. The truth is that you may have missed the boat on a block of instruction; your Cadre may not know or may not tell you the correct information before attending LDAC but in the end, it is your responsibility to look out for your best interests. When I joined the Army, I thought there was a large secret warehouse with equipment especially for war. After a few years, I realized this was not the case. In the same respect, there is not a large secret warehouse with the insights and knowledge that will be issued to you before attending LDAC. Therefore, you have to set yourself up for success by understanding that LDAC IS NOT A GAME, but a reality! Therefore, the first two Cold Hard Facts about LDAC are:

### COLD HARD FACT 1:

There is no secret warehouse that has all the knowledge dedicated to Cadets before they attend LDAC.

### COLD HARD FACT 2:

Cadets go to LDAC with the knowledge and insights that they have obtained since being affiliated with Army ROTC.

Since LDAC is the definitive test that will set the tone for the rest of a Cadet's career in the Army there are several Flat Ass Rules (FLARs) that must be adhered to without any deviations. For example, a FLAR for a skydiver is to wear a parachute in order to survive a jump. An example of a Cadet FLAR is to be at training and on time! While working at the University of Central Florida, LTC (R) Christensen introduced me to the acronym FLAR and over the next few chapters, several FLARs will be addressed. A known fact is that if Cadets fail to prepare for LDAC, they will fail! Therefore, it is imperative that Cadets take every advantage of every training event or aid that can assist them in succeeding. For that reason, I am giving you your first FLAR to survive LDAC and Accessions. Manage your career starting RIGHT NOW! Don't rely on someone else to take care of you during your military career. Start now and set yourself up for success by learning the ins and outs of LDAC, resulting in an increased awareness of your Operational Environment and understanding what to expect once you arrive at

LDAC. This book covers the main topics of LDAC and if used correctly can be a great guide to looking out for your best interest. This book answers the "How" and the "Why" to questions that Cadets have to include questions they did not even think about while in ROTC. When reading this book you will experience a transformation of the way you understand the "Why"behind LDAC. Although you will receive several examples in the book, it will ultimately be your responsibility to do well at LDAC.

As a future leader in the Army, you will be required to operate in an asymmetrical environment and conduct Full Spectrum Operations that will require creative thinking instead of adhering to a specific checklist or a trained situation. Trust me, you will experience situations with very little information that you never thought were possible or ever imagined; however, they will require a decision even though you may not have all the facts. Therefore, you must train to solve problems, train to increase intangible attributes (such as initiative, judgment, awareness, and confidence), train to increase your understanding and awareness, train to increase deliberate thought under stress and train to improve your unit's combat performance. Your focus must be on the total outcome of the task and not just the execution of the task. Therefore, it is imperative that you understand there is no specific template that will get you an "E" at the commissioning course. You must EARN it by understanding how to operate in an ever changing environment.

Finally, one of my FLARs is, "Don't start something that you don't finish!" With that said, it may have taken me several months to finish the book, but as stated before, "Come hell or high water, I will complete it and it will succeed."

Enjoy!

# How to Read and Use This Book

There is no doubt that you will be very busy during the MS III Year! For this book to have value and provide you with useful insight and knowledge to LDAC, you must read this book in its entirety multiple times and read specific sections again prior to and after attending class and lab. For example, if you read the Land Navigation section prior to attending class, you will be able to ask pertinent questions for future application. Reading it again after class will allow you to absorb more information and the ability to apply the information during practical exercises at lab. The sections are designed to be read relatively quickly and without a lot of fluff! The book assumes that you have a basic level of understanding and can apply the techniques and concepts with practice. If you are a MS I or II reading this book, it is designed to give you a head start and get you prepared for your MS III year. If you are a MS IV, the accessions process is outlined in detail to provide a Cadet the ability to war game for their branch of choice. Therefore, I encourage everyone who reads this book to adopt an active learning approach. Simply reading the text once will not give you the fidelity and level of understanding required for attending LDAC.

There will be items that do not make sense to you during the first reading. There will be areas that don't make sense to you after the second reading. However, if you apply yourself and create discussion during class and lab, there will be a point when you will experience a EUREKA moment. At this point, your vision will become clear about the insights and the necessity of these tidbits of information. Hold study groups with your classmates and discuss the information after reading information in this book, implement these concepts during labs and FTXs and most of all don't wait to implement this valuable information until you have read the entire book. Experiment with the ideas presented, reflect on these experiences and refine them before attending LDAC. Practice makes perfect. If you wait until two weeks before LDAC to read this book, you missed the boat so start reading it now! While reading this book or during practical application, a variation of the techniques may surface that suits your style of understanding and level of performance. By all means, if it works, implement it!

Finally, realize that you must study ROTC material just like information from any other college level course. I always demanded the best from Cadets and I never let them have the satisfaction of turning in a substandard assignment without going back and correcting their mistakes. You should demand this caliber of work from yourself. After all, there is a National Order of Merit List (OML) that places you in competition with your peers to receive your branch of choice. Therefore, study the material and increase your chances of receiving your desired branch. As with any article, book or instruction manual, there is no one secret that you can read to make you succeed at LDAC. However, I can assure you that reading this book will give you the insights, knowledge, and perspective to improve your skills, as well as the awareness of the Cold Hard Facts about LDAC, resulting in reducing your fear of LDAC.

Rodney Cruce
*Orlando, FL*

# Pre-LDAC

## Understanding the Situation

# 1 Identifying the Threat

Enemy Forces at Leadership Development and Assessment Course (LDAC) consist of Tiny Heart Syndrome (THS), Evaluators, Opposing Forces (OPFOR) and The Inner Villain (TIV). Every one of these enemy forces can be defeated with the proper preparation by ensuring you know their composition, disposition, capabilities and strength. From this information, you will deduct the Enemy Most Dangerous Course of Action (COA) and Enemy Most Likely Course of Action. If you know what the enemy's COAs are, then you can develop a plan to defeat the enemy. The final question that must be answered from this section of the book is the "So what?" The "So what?" answers the question "What does it mean to your unit and how it will affect the unit's mission?" The "So what?" for LDAC is that you must prepare to defeat the enemy and get LDAC Credit at all costs. LDAC Credit means that you have passed this gateway to get commissioned.

Many Cadets experience ***"Tiny Heart Syndrome"*** at one point or another during their ROTC career. Tiny Heart Syndrome is defined as "cowardice; reluctance to fight or to fully commit (to an undertaking or challenge)"; basically feeling sorry for one's self. THS is very difficult to recognize since it wears no specific uniform, has no equipment and does not travel in a formation that is easily identified. THS can be a dangerous enemy at LDAC and is capable of conducting psychological attacks with speed, surprise and violence of action. THS can be an overwhelming force that will destroy your motivation and will to lead. You do not want to be attacked by THS or be associated with this enemy at any cost! Your peers will see your inability to fight off THS as a weakness. Cadets who possess the intestinal fortitude to push through the tough situations prevail as true leaders by defeating THS. The best way to obtain the required intestinal fortitude to defeat THS is to train properly and display leadership at all times. Take advantage of every training opportunity that is given to you so that you are familiar with different situations. When in a leadership position, take charge and give good intent, guidance and direction. Always show the will to fight through the difficult times both physically and mentally—even when you're tired!

The ***"Elusive Evaluators"*** are not an enemy that will directly attack you at LDAC but they can affect your mission or overall evaluation. They are composed of both men and women who wear the ACU uniform with familiar patches that have the mobility to maneuver amongst Cadets with ease. These Evaluators come in many different shapes and sizes and will resemble Cadre at your university. At the platoon level, there are three Cadre Evaluators that are danger close and are responsible for evaluating your performance during Garrison and Forward Operating Base (FOB) evaluations. This three-person team is composed of a Platoon Train Advise Counsel (TAC) Officer known as a Platoon TAC Officer (PTO), a Platoon Train Advise Counsel Non-Commissioned Officer known as a PTNCO and a Platoon Train Advise Counsel Lieutenant known as a PTLT. At the Field Leadership Reactionary Course (FLRC), Squad Situational Training Exercises (SSTX) and Platoon Situational Training Exercises (PSTX), there is an unidentified number of Evaluators that will evaluate your performance during four of the six Cadet Leadership evaluations. Each Evaluator's objective is to give each Cadet a fair evaluation of his or her leadership skills in leadership positions at LDAC. The Evaluators are at 100% strength and have high morale. They see everything and have the capability to give Cadets an Excellence "E", Satisfactory "S" or Needs Improvement "N" Rating during Cadet leadership evaluations or Spot Reports. The Evaluators can be frightening and standoffish at times but they can be defeated by providing excellent leadership skills during missions and leadership positions. Cadets must not mistake this enemy as allies! The Evaluators may seem friendly and non-threatening in nature but they can become very deadly by employing observation skills and placing these observations on your evaluation. Once again, the best way to defeat this non-threatening enemy is by displaying outstanding leadership. Outstanding leadership will overwhelm the Evaluators and place the odds in your favor. Dedication to learning land navigation, physical fitness, regulations, field manuals, battle drills and squad and platoon operations will give you the necessary tools to defeat the Evaluators. Remember; don't let your guard down! You and your actions are always under close observation from the first moment you sign into LDAC until you are on the plane returning home!

***LDAC Opposition Forces (OPFOR)*** will be used as a dedicated enemy against your squad or platoon in the field phase of LDAC. There is an Annex B that is produced each year that explains the composition, disposition, capabilities and strength of the OPFOR. Most Cadets do not read this until they arrive at LDAC and by then it is too late because time becomes a scarce resource. Ask your Cadre for a copy of Annex B early in the spring semester. If they tell you they don't have it, be persistent and continue to ask for it until you receive it. Also, check the Warrior Forge website for updated information. Here is a little bit of information to get you started to understand the scenario at LDAC.

The OPFOR represents enemy insurgent forces known as South Atropian People's Army (SAPA). They are well trained and very familiar with the region along the Atropia and Ariana border. They usually operate in two to three person teams and use various machine guns, RPG-7 (Rocket Propelled Grenade) and assault rifles. Their higher headquarters control and coordinate these teams to synchronize attacks and mutually support each other's operations for long periods of time. Their primary mission is to destabilize local government through intimidation. SAPA teams are capable of conducting reconnaissance, small unit tactics (SUT) consisting of raids, ambushes and illegal check points. Other operations consist of terrorist attacks, emplacement of Improvised Explosive Devices (IED), and recruitment operations. Terrorist attacks are usually against local leaders, ethnic Atroprians, government institutions and NATO forces if they have the advantage. Their uniform is a mixture of military and civilian clothes.

A key task a Cadet must decipher is to discern Combatants from Non-Combatants which include local police forces and civilians. Although it may seem like it at times, there are not 2000 OPFOR personnel at LDAC so do your homework understand the scenario before attending LDAC and don't shoot the hunters. Remember OPFOR will assist in evaluating your tactical skills while at field training. Finally, I recommend that you ask your Cadre for Annex B to find out about the remainder of the SAPA capabilities and possible Courses of Action.

The last enemy at LDAC is YOU, ***"The Inner Villain" (TIV)***! The enemy composition is one person that accompanies you at all times and may very well be your most dangerous enemy at LDAC since it has the capability to lay dormant for days on end and then with very little warning, arise with an overwhelming speed and force to destroy your mission success. Additional capabilities include lack of confidence, lack of leadership, lack of training, lack of knowledge or a lack of desire to become tomorrow's leader. Although this is your most dangerous enemy at LDAC, it is easily defeated. The best way to defeat the Inner Villain is the same as defeating the Elusive Evaluators; demonstrate excellent leadership during missions and leadership positions. One other variable that can impact a Cadet's performance and can be categorized as TIV is a Cadet's personality. For instance, if a Cadet is an introvert and not outspoken, it can impact his or her performance at LDAC and put those Cadets who want to be branched Combat Arms, Military Intelligence or other highly sought out branches into jobs that they do not want. For that reason, Cadets must make a self assessment and modify their behavior accordingly to become great leaders. Seeing oneself is very hard for anyone but in order to succeed, you must make alternations to your faults, and prepare yourself by attending all ROTC training events, learning every aspect of field manuals and battle drills, and ensuring that you gain the respect of your peers at LDAC. Your peers will make you or break you during leadership positions. Be genuine, always think of the team and never be a Spotlight Ranger; apply those core attributes and you will defeat this enemy with little effort.

# 2 Leveraging Assets and Alliances

Friendly Forces at LDAC consists of Leadership Skills (Be, Know and Do Concept), Cadre, Upper Classmates, Pesky Peers and Additional Enablers; all will have a specific role in how you will either be defeated or defeat your enemies at LDAC, so don't take them lightly and always foster good relationships with all. The section below, in order of priority, explains how you must leverage these friendly forces to your advantage to make your experience at LDAC memorable and gain LDAC Credit.

***Leadership Skills*** are the best friendly force that you can have on your side. These leadership skills are categorized into Seventeen Leadership Performance Indicators (LPI) that you will be evaluated on at LDAC during your month long experience at Fort Lewis, Washington. A good leader has certain fundamental qualities and characteristics that can make or break your performance at LDAC. Observed behavior is used to enable Evaluators to measure your leadership qualities and characteristics by defining the appropriate Seventeen Leadership Performance Indicators (LPI). In Chapter 3, Leadership Performance Indicators from the U.S. Army Cadet Command Leadership Development Program (LDP) will be discussed to educate Cadets how they relate to the Seventeen Leadership Performance Indicators (LPI) and what type of rating would be given for that action.

ROTC ***Cadre*** will be a BIG influence and one of your best allies for maturity and professional development while attending ROTC both before and after LDAC. As a MS III Cadet, you must take advantage of the Cadre experience at your school to gain as much information as you can. You should be a sponge and not a rock! If you act like a sponge, you will absorb a significant amount of information. If you take the roll of a rock, you will repel the information and not have an extensive knowledge base to effectively apply good leadership. Although not all of your Cadre will have the same experience or knowledge, they all have something to offer that will ensure you are prepared and armed for LDAC; think of them as the secret warehouse in the sky. From personal experience as an ROTC Instructor, I can confirm that not all Cadre is the same caliber which will be very apparent as you enter the MS III year or even before if you have situational awareness by paying attention to your surroundings. Rest assured, there is at least one Cadre out there that can help you prepare for LDAC. Finally, don't negate the fact that other Cadre from different schools can provide crucial information during Field Training Exercises (FTX) and other training opportunities.

Your ***Upper Classmates*** can provide some of the best and worst information before attending LDAC. You have to sift through the crap and determine what Cadets have the best information and your interest at heart. Ask your MS III Instructor which upper classmates to align with so that you can benefit from their LDAC experience. This is a situation where you have to decide whether to be a sponge or a rock. For example, if the information from an upper classmate seems a bit far fetched or not in line with Cadre instruction, be a rock with appropriate courtesy and verify to determine who has the correct information. All your MS IVs will act like they have the solution to LDAC. They will refer to their experiences and make the situation seem to be worse than what it is or what it will be the next year. So, don't get spooked! Welcome the challenge and realize that you control your own destiny. After all they made it through LDAC and you are much better than they ever thought about being. After you get through that first MS III semester, you will see perfectly with 20/20 vision which MS IVs can provide the most reliable, and creditable insights and knowledge before attending LDAC. Don't be afraid to ask the MS IVs what rating they received at LDAC to include what techniques they used to succeed. This will assist you in determining who knows what he or she is talking about. Although an individual who received an Excellence "E" rating at LDAC usually has a good understanding of what must be accomplished to receive a successful rating, you must realize that some of them were very lucky. The best advice that I can recommend is to use common sense and use multiple sources to gain as much information as you can.

Those ***"Pesky Peers"*** can make and break your leadership evaluations both on campus and at LDAC. Prior to attending LDAC, your peers play an important and active role in developing your leadership skills and most of all they are an unlimited resource during missions if you have the ability to motivate and empower them when in a leadership position. One method of getting your peers to work for you on campus that has shown great success is a ROTC study group. I recommend that one night a week you and your peers meet at someone's house and have a study group. All MS III Cadets have a different level of knowledge and experience that can help each other learn new things. One example is to practice issuing Operations Orders (OPORD) to each other, discuss how you would make leadership decisions for different scenarios. Another example is to practice technical skills such as route planning during land navigation. Your scope is limitless and will only help you in the end. While at LDAC, your peers can assist you in getting a great leadership evaluation but you must treat them with respect and be genuine in your actions. At LDAC you will receive one evaluation at the squad level and the other at the platoon level. Although peer evaluations are not a Go or No Go event, they can push you over the edge to get that Excellence Rating or be named as the Platoon Top 5. As a previous Platoon TAC at LDAC, confidence and genuine actions go a long way if you treat your peers as equals and do not get a big head while in charge. Think of it as a contract between you and your peers so that when they are in leadership you support them and when you are in leadership, they support you. Finally, the worst thing that you can do is turn your peers into enemies (Refer to paragraph Enemy Forces; The Inner Villain).

***Additional Enablers*** consist of Field Manuals, TACSOP, FTXs, prior experience and training that assist Cadets in receiving LDAC credit. Each Cadet will have different experiences and will be at different skill levels before arriving at LDAC. Therefore, Cadets must understand what their strengths and their weaknesses are so they can form alliances with the best enablers to assist them. Listed below are the minimum enablers that are needed to give Cadets the fighting chance to ensure they are prepared for the rigor that is induced at LDAC.

*Field Manuals* contain useful information that educates you on procedures, concepts and guidelines that will facilitate decision making at LDAC. Cadets who read these regulations do much better than Cadets who rely only on class information. ROTC classes are designed to introduce and familiarize Cadets with concepts and labs are designed to allow for practical exercises. All of your questions will not be answered in class or at lab, therefore, Cadets must act like college students and read their manuals. The difference between Field Manuals and your class text books is that there are multiple Field Manuals as opposed to just one text book for the class. The good thing is that most of the manuals have plenty of pictures. I recommend reading field manuals a minimum of two times to include taking notes and studying what you read. If you read the proper field manuals that cover a certain subject prior to attending your ROTC class, you will be quite surprised at the increase in knowledge you will have and how much more you understand. To assist you, we offer the Ranger Hand Book on our website at www.myleadersource.com cheaper than anyone else on the market. Additionally, the following manuals have pertinent inform ation that will assist you at LDAC:

| | |
|---|---|
| FM 3-21.8 | The Infantry Rifle Platoon and Squad |
| FM 3-21.10 | The Infantry Rifle Company |
| FM 3-21.75 | The Warrior Ethos and Soldier Combat Skills |
| FM 3-21.5 | Drill and Ceremonies |
| FM 3-22.9 | Rifle Marksmanship, M16/M4 Series Weapons |
| FM 3-25.26 | Map Reading and Land Navigation (entire Field Manual) |
| FM 7-8 | Company and Platoon Infantry Manual |
| STP 21-1-SMCT | Soldier Manual of Common Tasks |
| FM 21-18 | Foot Marches |
| FM 21-20 | Physical Fitness Training |
| SH 21-76 | Ranger Hand Book |

***"The Cold Hard Facts About LDAC"*** is a great base of knowledge from an Instructor and Evaluator's viewpoint and will direct your focus in preparing for LDAC. This is a proven template that has developed numerous MS III Cadets for attending LDAC with tremendous success. This success ranges from being #1 in the regiment to receiving a Needs Improvement "N" at LDAC. Different Cadets need different methods for learning. If used correctly, this book will be a roadmap and guide to informing you how to train, what questions to ask and where to focus your efforts in order to prepare you for LDAC.

***Tactical Standard Operating Procedure (TACSOP)*** is the life line that is provided to every Cadet and is allowed at LDAC. The TACSOP should be treated as a sensitive item and used extensively for every leadership evaluation. A rule of thumb is to reference the TACSOP prior to assuming a leadership position so that you will remain on azimuth. In the section Troop Leading Procedures, there is an Assumption of Command section that has a checklist of what you should do when taking over a leadership position. This is just one example of how to use the TACSOP but in order to maximize its potential; you must become very familiar with how to reference it without calling a friend. Although the TACSOP is put together fairly well, you will have to experiment with what works for you before attending LDAC. Finally, there are a few modifications that you should make to the TACSOP before arriving at Fort Lewis. These changes will assist you at different times during your month long adventure since you can't reference any other Field Manuals or references materials. Now, make the following additions to sections in the TACSOP and don't lose it.

1) Write your name in the front cover with a Sharpie.

2) In the Orders Section add the words Visual Aids and list the visual aids that you will need to construct before giving out your OPORD in Garrison. It is very important that you have these 4 visual aids at a minimum. Use them when issuing out your OPORD and you should receive positive marks in the Leadership Performance Indicators (LPI) of IN, CO and DK. It is crucial to ensure that you delegate who will be responsible for each of the visual aids so that all of your time is not wasted. One key note is that CO is the most evaluated leadership dimension at LDAC.

## Visual Aids

Tasks – PL

Graphics – PL

Uniform – PSG

Timeline – PSG

3) In the Orders Section add notes for the FLRC event. These notes will remind you to think of FLRC as a mission instead of a fun team building task. The model below reminds you how to use the 4 leadership dimensions that are known as the Golden Diamond and how to ensure you are on the right track for a good evaluation. You will learn more about this concept in Chapter 8.

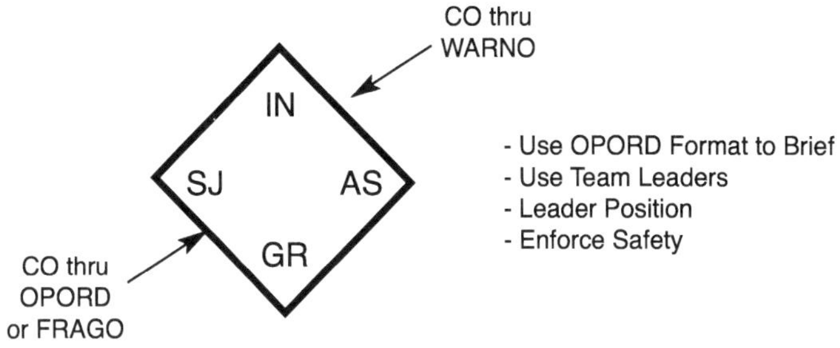


**Figure 1: Golden Diamond**

4) In the Reports Section write in the Dos and Don'ts. DOs that will get you positive marks and DON'Ts can hurt you on the peer ratings and the communication dimension.

| DO | DON'T |
|---|---|
| Orchestrate the AAR | Don't be the only one talking |
| Get SL/TMs involved | Don't say the word "I" |
| Get other members involved | Don't incriminate yourself |
| Ask a lot of questions | Don't brag about yourself |
| Talk in Phases & be positive | Don't single anyone out |
| Keep the AAR 15 minutes or less | |

# 3 The Leadership Development Program

**The Leadership Development Program (LDP)** is a multiple step process used at LDAC during all training events to elicit and assess the leadership potential of ROTC Cadets with instant feedback. The process is individually focused with a standard assessment that measures behavior during six different leadership opportunities with instant feedback for leader development and is automatically linked to the accessions process for efficiency and timely feedback in order to meet the September deadline for branching assignments. The assessment hinges on recognizing, recording, classifying and rating a Cadet's behavior on the overall effectiveness and what you would reasonably expect from a MS III Cadet. In reality, the LDP is very efficient and allows for a fair evaluation. However, from a Cadet's view point, there are two perceptions to how the process works: the Cadet Perspective and everyone else. Over the next few pages, the LDP Process will be discussed to educate Cadets on how it actually works using the Army's BE-KNOW-DO concept to express what attributes are required of Army leaders, the LDP seeks to develop cadets into agile, multi-skilled officers.

Recognizing behavior is the first step of the LDP process. Evaluators consist of Cadre from ROTC programs across the U.S. that are trained and calibrated during a three-day course to ensure they understand the difference between critical, non-critical, physical and mental behavior. The major difference between critical and non-critical behavior is that critical behavior has a significant impact on a leader's effectiveness where non-critical has little if any impact on the Cadet's effectiveness. Cadets' physical and mental behavior must be analyzed to determine an overall assessment of their actions that relate to the Seventeen Leadership Performance Indicators (PLI). Evaluators monitor Cadet behavior by firsthand or secondhand observations. If an evaluator observes the behavior firsthand, the result is usually noted because the Cadre is physically at the location where the behavior occurred. The secondhand observation is much more important because it uses deduction to gather the information. Many Cadets fail to understand that just because a Cadre did not physically observe their actions, he or she did document the behavior. Secondhand observations are gained through results, other Evaluators or asking certain Cadets specific questions. I used this technique to determine where the communication breakdown occurred when a task was not completed. Since it is very hard to measure intellectual capacity leadership behaviors outright, the secondhand observation becomes a useful tool to the Evaluator but the he or she must judge the accuracy of the source before coming to a conclusion. Therefore, if an Evaluator is asking you a question, don't get nervous because he or she is investigating what critical thinking took place, guiding you in the right direction or determining which Cadet in the chain of command did not follow the eight Troop Leader Procedures. During the Leadership Development Process, leadership behavior is observed and documented by a Blue/Green Card while Cadets are in leadership positions or by a Spot Report when they are followers. When Cadets are in leadership positions, there are scheduled situations that create behavior that can be observed regularly. The orders process is used in great detail to create recurring behavior so that Evaluators can make honest assessments of the leaders' actions. For followers, Spot Reports provide observation opportunities that are non-reoccurring such as initiative, conflicts and personal interaction. Finally, in my opinion, the Spot Report provides more information as to what Cadets' true character and leadership ability are because they do not feel they are being observed since they are out of a leadership position; therefore don't let your guard down.

Recording behavior is the second step of the LDP. Evaluators must record what behavior occurred during the situation. Since Evaluators usually have two Cadets to evaluate, they either write the actions down or make a mental note. When I was a PTO, I took notes on each Cadet that I evaluated and believe it or not, I still have those notes today. I dedicated two sheets of paper to each Cadet during his or her leadership evaluation so that I did not have to search multiple locations when it was time to write the evaluation. I only wrote down what I thought was critical behavior. Outside that critical behavior I documented what leadership behavior (known as Classification) was experienced to include if it was positive or negative. Some Cadre annotates everything they observe and then decipher the results at the end. Other Cadre watch, followed by a summary of leader's actions in bullet form to support the leadership behavior observed. Many different leadership behaviors can be observed through one single action. If this is the case, the write up should encompass each of those leadership behaviors. Remember, each Evaluator's capability will be different therefore their method of documentation will vary so don't think that one method is better than the next.

Next a Cadet's behavior must be Classified. The Army's Leadership Requirements Model encompasses two major leadership sections: Attributes and Core Leadership Competencies. The first section, "Attributes", are "What a Leader Is" and focuses on three areas that assist the Evaluator to classify behavior: 1) A Leader of Character 2) A Leader with Presence 3) A Leader with Intellectual Capability. A very critical piece of information that Cadets need to understand is that A Leader of Character (The seven Army Values, Empathy and Warrior Ethos), are attributes that refer to the minimum standards for a Cadet to be a member of the U.S. Army. These beliefs cannot be sacrificed at any cost. If a Cadet receives a "No" in any of these attributes, they will have a very poor rating which could result in the Cadet attending a board that does not give the Cadet credit for LDAC. The second section of the model is "Core Leadership Competencies". This part focuses on "What a Leader Does" and uses three areas in assisting Evaluators to classify behavior: 1) Leads 2) Develops and 3) Achieves.

The seven Army Values, Empathy, Warrior Ethos and seventeen leadership behaviors are the driving force to Evaluator Comments to include how they are classified. This is a very important section because if you understand it, you will be able to have the emotional intelligence to recognize leadership behaviors in your peers and determine which ones you are lacking during your LDAC experience. Each of these areas and leadership behaviors are covered in intimate detail below so that a level of understanding is attained about what each and every one of them signifies.

The Army Values are the first area of evaluation that is learned by Cadets during their first year at ROTC and is easily remembered by the acronym LDRSHIP. These seven Army Values are leadership behaviors and are expected from Cadets so they will not be annotated on a Blue/Green card unless an individual violates one of these values. The Army's Leadership Manual (FM 6-22) describes the model in extensive detail but I will summarize the information of what a Cadet needs to understand before attending LDAC:

## Army Leadership Requirements Model

Section 1: Attributes (What an Army Leader Is) - Characteristics that are an inherent part of an individual's total core, physical, and intellectual aspects. Attributes shape how one behaves

1) A Leader of Character (BE) - A person's moral and ethical qualities which give a leader motivation to do what is appropriate regardless of circumstances or consequences

Army Values – Loyalty (LO), Duty (DU), Respect (RE), Selfless Service (SS), Honor (HO), Integrity (IT) and Personal Courage (PC).

Empathy (EP) - The ability to see something from another person's point of view, to identify with and enter into another person's feelings and emotions

Warrior Ethos (WE) - I will always place the mission first; I will never accept defeat; I will never quit; I will never leave a fallen comrade

2) A Leader with Presence (KNOW) - The impression that a leader makes on others, which contributes to their success in leading them; the image that a leader projects; how others perceive a leader (outward appearance, demeanor, words, and actions). This attribute measures four leadership behaviors to assess Cadets: Military Bearing (MB), Physical Fit (PF), Confident (CF) and Resilient (RS).

3) A Leader with Intellectual Capacity (KNOW) - The ability to draw on the mental tendencies and resources that shape a leader's conceptual abilities and impact of effectiveness, which then are applied to one's duties and responsibilities This attribute measures five leadership behaviors to assess Cadets: Mental Agility (MA), Sound Judgment (SJ), Innovation (IN), Interpersonal Tact (IP) and Domain Knowledge (DK).

Section II: Core Leader Competencies (DO) - Work to lead others; develops themselves, their subordinates and organizations to achieve mission. The three areas are:

1) Leads - The application of character, presence, intellect and abilities while guiding others toward a common goal and mission accomplishment. This Core Leadership Competency measures four leadership behaviors to assess Cadets: Leads Others (LD), Extends Influence beyond Chain of Command (CoC), Leads by Example (LE) and Communicates (CO).

2) Develops - Taking actions to foster team work, encourage initiative, and to accept personal responsibility, while demonstrating care. These Core Leadership Competency measures three leadership behaviors to assess Cadets: Creates a Positive Environment (CP), Prepares Self (PS) and Develops Others (DO).

3) Achieves - Sets objectives and focuses on mission accomplishment. Although this only measures one leadership behavior, Gets Result (GR), many different types of critical behavior can be evaluated.

**Leadership Performance Indicators**

The final step is to Rate the behavior which results in the base net assessment on how effective the Evaluator thought the

## Table 1 Leadership Performance Indicators (LPI)

| Section | Army Model Leadership Selections | Leadership Performance Indicators (LPI) or Leadership Behaviors/ Dimensions |
|---|---|---|
| I | **Attributes:** What an Army Leader Is | |
| | A Leader of Character | LO, DU, RE, SS, HO, IT, PC, EP, WE |
| | A Leader with Presence | MB, PF, CF, RS |
| | A Leader wtih Intellectual Capacity | MA, SJ, IN , IP, DK |
| II | **Core Leadership Competencies:** What An Army Leader Does | |
| | Leads | LC, CoC, LE, CO |
| | Develops | CP, PS, DO |
| | Achieves | GR |

Cadet's leadership went overall. If a Cadet receives an Excellent "E" Rating, then the leader's actions were above what a Cadre would reasonably expect from a MS III Cadet. When rating the actions for an Excellence leadership behavior, the observation executed should be positive. A Satisfactory "S" Rating is what an Evaluator would expect to see from a MS III Cadet. This rating allows for some mistakes and is given if the observation executed was generally neutral. A Needs Improvement "N" Rating is below what you reasonably expect from a MS III Cadet. As an example, the leader may demonstrate effective emotional control throughout most of the assessment, but a single, momentary lapse of self-control might negate all positive behavior exhibited throughout the day. Similarly, the Cadet may show little proficiency at the beginning of the rating period, only to show marked improvement near the end. The assessor may determine the improved performance more accurately reflects the leader's capabilities, and rate accordingly. One key note to understand is that just because you have more leadership behavioral Es than Ss or Ns, does not constitute that you will automatically receive an overall "E" Rating. After an Evaluator rates a Cadet's leadership position, the Cadet is then counseled using the Leadership Assessment Report known as Blue/Green Cards that address Attributes and Core Leader Competencies. During a leader's counseling, Cadre can view an individual's body language and reactions that can result in another observation so don't let your guard down by thinking the leadership rotation is over. While receiving the counseling, the Cadre should recommend to the Cadet how to improve their weak performance indicators. If they don't, ask them how you can improve for the next rating. This performance feedback is not a one way conversation. If you apply what the Cadre suggests, it could result in a higher rating on the following leadership evaluation. During the counseling, Cadets should track their progress by documenting each leadership behavior during the counseling. Continue to read the book and you will learn about the leadership behavior tracker that can assist Cadets at LDAC as you read this book.

The LDP is a powerful process that works if properly used. Understanding this process will allow Cadets to see the Cadre point of view. The best way to practice using this valuable information is to rate your peers during the MS III year. You and all of your peers will have leadership opportunities at home campus and with that comes an expectation of a certain level of proficiency and leadership from your peers. By placing yourself in the Evaluator shoes, you will want to see certain actions and if those actions are not observed, you will criticize them. Instead of criticizing other Cadets, make an evaluation of their leadership and recommend solutions on how they can improve without being a know it all or pompous ass. Ask them for recommendations so that you will automatically learn what must be shown at LDAC. The more you understand this process, the more you will benefit. This is a great opportunity to view this as an education on how you will view your Platoon Sergeant, Squad Leaders and soldiers once you become a Platoon Leader in the Army.

**Leadership Insight 1:** The Officer Evaluation Report (OER) is an annual evaluation that all Military Officers receive and it is based off similar leadership behaviors or Leadership Performance Indicators (LPI) used to evaluate Cadets at LDAC.

### COLD HARD FACT 3:

The LDP works and is a proven system!

***Satisfactory Versus Excellent Ratings Explained***

The difference between a satisfactory level of performance and an excellent level of performance can be a fine line since Evaluators have different expectations, prior experiences and pet peeves. If Evaluators are not on the same sheet of music or become biased, this can become an unfair evaluation process for the Cadets. LDAC has recognized this issue and to mitigate these factors, each Evaluator must attend a three day course with practical exercises to ensure they are calibrated before writing their first evaluation. Below are Leadership Performance Indicators or guidelines that Evaluators use to determine the difference between Excellent and Satisfactory levels of performance for the seventeen leadership behaviors to ensure Cadet's behavior is classified correctly. Look very closely at the difference between the two ratings because two Cadets may complete the same task but the difference may be how they accomplished the task. The words that outline the difference are italicized and most of all, understand that just completing a task does not check the block to receive an Excellent "E" evaluation. All other performance not meeting the satisfactory level is considered a Needs Improvement "N". Now take out a Blue Card for reference to assist you when reading each of the leadership behaviors listed below under their specified Attribute or Core Leadership Competency:

***A Leader with Presence***

What an Army Leader is and the impression that a leader makes on others, which contributes to their success in leading them; the image that a leader projects; how others perceive a leader (outward appearance, demeanor, words, and actions).

## 1) Military Bearing (MB)

**Projecting a commanding presence, a professional image of authority**

**MILITARY BEARING (MB)**

| EXCELLENT | SATISFACTORY |
|---|---|
| *EXEMPLARY* PHYSICAL AND PROFESSIONAL EXPERIENCE | *GOOD* PHYSICAL AND PROFESSIONAL APPEARANCE |
| *SETS EXAMPLE* FOR PHYSICAL FITNESS, MILITARY BEARING | MEETS STANDARDS FOR PHYSICAL FITNESS, MILITARY BEARING |
| DEMONSTRATES COMMAND PRESENCE, CADETS WANT TO FOLLOW | DEMONSTRATES *ADEQUATE* COMMAND PRESENCE |

## 2) **Physically Fit (PF)**

**Having sound health, strength, and endurance, which sustain emotional health and conceptual abilities under prolonged stress**

**PHYSICALLY FIT (PF)**

| **EXCELLENT** | **SATISFACTORY** |
|---|---|
| *REFUSES TO QUIT;* RELIABLY COMPLETES PHYSICALLY DEMANDING ENDEAVORS | PERSEVERES IN FACE OF ADVERSITY; *COMPLETES MOST* PHYSICALLY DEMANDING ENDEAVORS |
| *PROACTIVE* TOWARD OWN HEALTH; RECOVERS QUICKLY FROM PHYSICALLY DEMANDING EVENT, READY FOR NEXT | ATTENTIVE TO OWN HEALTH; SEEKS MEDICAL ASSISTANCE *WHEN NECESSARY;* FULFILLS TREATMENT DIRECTIVES |

## 3) **Confident (CF)**

**Projecting self-confidence and certainty in the unit's ability to succeed in whatever it does; able to demonstrate composure and outward calm through steady control over emotion.**

**CONFIDENT (CF)**

| **EXCELLENT** | **SATISFACTORY** |
|---|---|
| STRONG *SELF-CONFIDENCE* AND POSITIVE | *ADEQUATELY* SHOWS SELF-CONFIDENCE, POSITIVE |
| REMAINS CALM, UNDER CONTROL, EFFECTIVE UNDER PRESSURE; *NEVER COMPLAINS* | TRIES, WITH SOME SUCCESS, TO BE CALM AND EFFECTIVE UNDER PRESSURE, *RARELY COMPLAINS* |
| EXCELS BY STRESS, CHAOS AND RAPID CHANGE | FUNCTIONS UNDER STRESS OR AMID CHAOS AND RAPID CHANGE |
| BALANCES COMPETING DEMANDS | SUCCESSFULLY BALANCES MOST COMPETING DEMANDS |

## 4) Resilient (RS)

**Showing a tendency to recover quickly from set-backs, shock, injuries, adversity, and stress while maintaining a mission and organizational focus**

**RESILIENT (RS)**

| EXCELLENT | SATISFACTORY |
|---|---|
| RECOVERS *QUICKLY* FROM SETBACKS, SHOCK, INJURY, ADVERSITY AND STRESS WHILE MAINTAINING MISSION *FOCUS* | RECOVERS FROM SETBACKS, SHOCK, INJURY, ADVERSITY AND STRESS *WHILE MAINTAINING MISSION FOCUS* |

### Presence and Intellectual Capacity

The ability to draw on the mental tendencies and resources that shape a leader's conceptual abilities and impact of effectiveness, which then are applied to one's duties and responsibilities. Conceptual abilities enable sound judgment before implementing concepts and plans. They help one think creatively and reason analytically, critically, ethically, and with cultural sensitivity to consider unintended as well as intended consequences. Some decisions may set off a chain of events. Therefore, leaders must attempt to anticipate the second- and third-order effects of their actions. Even lower-level leaders' actions may have effects well beyond what they expect.

## 5) Mental Agility (MA)

### Flexibility of mind, a tendency to anticipate or adapt

**MENTAL AGILITY (MA)**

| **EXCELLENT** | **SATISFACTORY** |
|---|---|
| DEMONSTRATES *STRONG WILL, SELF-DISCIPLINE, CONFIDENCE, AND INTELLIGENCE* | SHOWS EVIDENCE OF *SOME WILL,* INITIATIVE, CONFIDENCE AND INTELLIGENCE |
| ANALYZES SITUATIONS; COMBINES COMPLEX IDEAS INTO FEASIBLE COURSES OF ACTION | *USUALLY* GENERATES REASONABLE COURSES OF ACTION THROUGH ANALYSIS OF SITUATIONS |
| DEMONSTRATES *CONSISTENT* COMMON SENSE | *USUALLY OR ADEQUATELY* SHOWS COMMON SENSE |
| *RECOGNIZES,* READILY ADOPTS BETTER IDEAS | SHOWS ABILITY TO ADOPT BETTER IDEAS |
| *RELIABLY* COMPLETES MENTALLY DEMANDING ENDEAVORS | *USUALLY* FINISHES DIFFICULT OR MENTALLY DEMANDING ENDEAVORS |
| SKILLFULLY HANDLES MULTIPLE DEMANDS | FACED WITH MULTIPLE DEMANDS, HANDLES *MOST* OF THEM EFFECTIVELY |
| *DYNAMIC SELF-STARTER,* ORIGINATES IDEAS AND ACTIONS | ORIGINATES *SOME IDEAS* OR ACTIONS |

# 6) Sound Judgment (SJ)

## The capacity to assess situation and draw feasible conclusions

**SOUND JUDGMENT (SJ)**

| EXCELLENT | SATISFACTORY |
|---|---|
| *DRIVES EVENTS* THROUGH SMART, TIMELY DECISIONS | MAKES DECISIONS; KEEPS PACE WITH EVENTS |
| *THOROUGHLY* ASSESSES SITUATION; ACTS ON KEY ISSUES AND LIKELY CONTINGENCIES/CONSEQUENCES | *ADEQUATELY* ASSESSES SITUATION; SORTS OUT IMPORTANT ASPECTS AND DECIDES ACCORDINGLY |
| *MAKES* THE "HARD, RIGHT DECISION" RATHER THAN THE "EASY WRONG" | *USUALLY* PUTS BEING RIGHT AHEAD OF BEING POPULAR OR EASY |
| BALANCES FIRM RESOLVE & FLEXIBILITY RECOGNIZES A BETTER IDEA AND INCORPORATES INTO OWN DECISION | *ATTEMPTS* TO BALANCE RESOLVE AND FLEXIBILITY; NOT AFRAID OF BETTER IDEAS FROM OTHER SOURCES |
| *RECOVERS QUICKLY* AFTER LEARNING A DECISION WAS INCORRECT | *PAUSES*, BUT REMAINS DECISIVE AFTER LEARNING A DECISION WAS INCORRECT |
| *QUICKLY* RECOGNIZES NEED AND DECIDES/ACTS IN THE ABSENCE OF GUIDANCE | *EVENTUALLY* RECOGNIZES NEED TO DECIDE/ACT IN THE ABSENCE OF GUIDANCE AND DOES SO |
| *THOROUGHLY* CONSIDERS PROBABLE IMPACTS AND CONSEQUENCES OF OWN DECISIONS | ADEQUATELY CONSIDERS PROBABLE IMPACTS AND CONSEQUENCES OF OWN DECISIONS |

## 7) Innovation (IN)

### Ability to introduce something new

**INNOVATION (IN)**

| EXCELLENT | SATISFACTORY |
|---|---|
| *STRONG JUDGMENT;* CRITICAL/ CREATIVE THINKING AND REASONING | *GENERALLY* SHOWS GOOD JUDGMENT, SOUND THINKING AND REASONING |
| *ANTICIPATES* REQUIREMENTS OR CONTINGENCIES | READILY REACTS TO REQUIREMENTS AND CONTINGENCIES |
| *IMPROVISES* WITHIN COMMANDER'S INTENT | READILY UNDERSTANDS AND FOLLOWS COMMANDERS' INTENT |
| CAN ANALYZE/SYNTHESIZE | SHOWS *SOME* ANALYTICAL AND SYNTHESIZING SKILLS |
| *COMFORTABLE* IN REALM OF IDEAS; INNOVATIVE | CAN FUNCTION IN REALM OF IDEAS |

## 8) Interpersonal Tact (IP)

### Effectively interacting with others by understanding others perspectives

**INTERPERSONAL TACT (IP)**

| EXCELLENT | SATISFACTORY |
|---|---|
| *READILY* INTERACTS WITH OTHERS; EARNS RESPECT | *ADEQUATELY* INTERACTS WITH OTHERS |
| LEVERAGES SKILLS AND ABILITIES OF TEAM MEMBERS | GETS ALONG WELL WITH OTHERS |

| | |
|---|---|
| ACTIVELY CONTRIBUTES TO PROBLEM-SOLVING AND DECISION MAKING PROCESS | *CONTRIBUTES, WHEN ASKED,* TO PROBLEM SOLVING AND DECISION MAKING PROCESSES |
| *SOUGHT OUT* BY PEERS FOR EXPERTISE OR COUNSEL | *SOMETIMES* ASKED FOR COUNSEL BY PEERS |

## 9) Domain Knowledge (DK)

### Possessing facts, beliefs, and logical assumptions in many areas

**DOMAIN KNOWLEDGE (DK)**

| EXCELLENT | SATISFACTORY |
|---|---|
| *THOROUGH KNOWLEDGE* OF TASK ACCOMPLISHMENT STANDARDS | *WORKING KNOWLEDGE* OF STANDARDS FOR TASK PERFORMANCE |
| *MASTERS* DUTY TASKS | *STRIVES FOR* MASTERY OF DUTY PERFORMANCE |
| *STRONG GRASP* OF BASIC TACTICS, TECHNIQUES, PROCEDURES (TTP) | *UNDERSTANDS* BASIC TACTICS, TECHNIQUES, PROCEDURES |
| SKILLED IN PREPARATION OF OPERATIONS ORDERS | PREPARES ADEQUATE OPERATIONS ORDERS |
| *EXPERT* IN BASIC SOLDIER SKILLS | *COMPETENT* IN BASIC SOLDIER SKILLS |
| EFFECTIVELY APPLIES SKILLS TO TRAIN FOR OR TO FIGHT AND WIN WARS | HAS BASIC SKILLS TO TRAIN FOR OR TO FIGHT AND WIN WARS |

***Core Leader Competencies*** – What an Army Leader Does. As a leader your ACTIONS are louder than your words. When in leadership, you have to be able to "walk the talk" or your subordinates will not have the confidence in your abilities to lead them through difficult situations. There are eight leadership behaviors that are divided into three categories; Leading, Developing and Achieving: The core leader competencies emphasize the roles, functions, and activities of what leaders do. Leaders serve to provide purpose, direction and motivation. LDP strives to develop Army leaders work hard to lead people, to develop themselves, their subordinates, organizations, and to achieve mission accomplishment across the spectrum of conflicts.

***A. Leads*** - The influencing of character, presence, intellect and abilities while guiding others toward a common goal and mission accomplishment.

# 10) Leads Others (LD)

## Influencing Soldiers or Army Civilians in the unit or organization

**LEADS OTHERS (LD)**

| EXCELLENT | SATISFACTORY |
|---|---|
| *CLEARLY AND POSITIVELY* ARTICULATES EXPECTATIONS | *ADEQUATELY* DEFINES REQUIREMENTS AND GUIDANCE |
| INSPIRES ACTION BY APPROPRIATELY ACKNOWLEDGE GOOD PERFORMANCE AND DEALING WITH POOR PERFORMANCE | ACKNOWLEDGES GOOD AND POOR PERFORMANCES; TAKES SOME ACTION ACCORDINGLY |
| *SKILLFULLY* DELEGATES CONSISTENT WITH REQUIREMENTS OF DUTY POSITION AND INDIVIDUAL'S CAPABILITY AND DEVELOPMENTAL NEEDS | GIVES THOUGHT TO DUTY POSITION, CAPABILITY, AND DEVELOPMENTAL NEEDS WHEN DELEGATING |
| MEETS SUBORDINATES' NEEDS, KEEPS THEM INFORMED, PROVIDES RATIONALE AND PROVIDES EARLY WARNING | *ATTEMPTS* TO MEET SUBORDINATE'S NEEDS, KEEP THEM INFORMED, PROVIDE RATIONALE AND PROVIDE EARLY WARNING |
| *ACTIVELY* LISTENS, SEEKS FEEDBACK AND MAKES SMART, TIMELY ADJUSTMENTS TO ACTIONS OR TASKINGS | MAKES SOME USE OF FEEDBACK TO MODIFY ACTIONS AND TASKINGS IN PROGRESS WHEN NEEDED |

## 11) **Extends Influence Beyond CoC (EI)**

### Ability to operate in an environment encompassing higher and lower command structures

**Extends Influence beyond CoC (EI)**

| EXCELLENT | SATISFACTORY |
|---|---|
| *CONSISTENTLY* FOLLOWS THROUGH ON ACTIONS RELATED TO EXPECTATIONS OF OTHERS | FOLLOWS THROUGH ON ACTIONS RELATED TO EXPECTATIONS OF OTHERS |
| EASILY GAINS COOPERATION OR SUPPORT WHEN WORKING WITH OTHERS | *ATTEMPTS TO* GAIN COOPERATION OR SUPPORT WHEN WORKING WITH OTHERS |
| EASILY INFLUENCES PERCEPTIONS ABOUT THE UNIT OR ORGANIZATION | *MAKES SOME* INFLUENCE ON THE PERCEPTION |

## 12) **Leads by Example (LE)**

### Providing an example that others consider and use in what they do

**LEADS BY EXAMPLE (LE)**

| EXCELLENT | SATISFACTORY |
|---|---|
| *QUICKLY* ADAPTS TO CHANGING ENVIRONMENTS AND SITUATIONS | ADAPTS TO CHANGING ENVIRONMENTS AND SITUATIONS *OVER TIME* |
| CONTINUOUSLY EXERTS LEADERSHIP BY PERSONAL PRESENCE AND EXAMPLE | EXERTS LEADERSHIP BY PERSONAL PRESENCE AND EXAMPLE |

| | |
|---|---|
| *QUICKLY* APPLIES CREATIVE THINKING IN IN COLLABORATION WITH SUBORDINATE LEADERS | APPLIES CREATIVE THINKING IN COLLABORATION WITH SUBORDINATE LEADERS |

## 13) Communicates (CO)

### Attain a clear understanding of what needs to be done and why within / without their organization

**COMMUNICATES (CO)**

**ORAL**

| EXCELLENT | SATISFACTORY |
|---|---|
| *CONVEYS* IDEAS AND FEELINGS CONCISELY IN A SINGLE TRANSMISSION | *GENERALLY* CONVEYS IDEAS, FEELINGS, SINCERITY |
| EXPRESSIONS *WELL-PLANNED,* ORGANIZED AND SUITED TO AUDIENCE AND SITUATION; INSPIRES OWNERSHIP, ATTENTION AND CONVICTION | EXPRESSIONS USUALLY WELL CONCEIVED WITH EVIDENCE OF SOME ORGANIZATION; GENERALLY INTERESTING AND CONVINCING |
| *RECOGNIZES* MISUNDERSTANDINGS AND RESOLVES THEM CLEARLY IN A POSITIVE, TEAM-BUILDING MANNER | ACTS TO DETERMINE/RECOGNIZE MISUNDERSTANDINGS AND RESOLVES THEM TO LARGE EXTENT |
| CONVINCES AND/OR PERSUADES UNIT; BUILDS OWNERSHIP OF PLAN | KEEPS AUDIENCE ON TRACK |
| FREE FROM GRAMMATICAL ERRORS—TERMS, PHRASES ALWAYS FIT THE AUDIENCE | *USUALLY* USES GRAMMATICALLY CORRECT TERMS AND PHRASES |

| | |
|---|---|
| CLEAR, CONCISE VISUAL AIDS; SMOOTH PRESENTATION; USES INFLECTION TO ADVANTAGE | *ADEQUATE* VISUAL AIDS FEW DISTRACTING GESTURES OR SOUNDS; MAKES ADEQUATE USE OF INFLECTION |
| ATTENTIVE, SUPPORTIVE LISTENER; MAKES APPROPRIATE NOTES; GOOD RETENTION OF CRITICAL DETAILS | LISTENS AND WATCHES ATTENTIVELY; MAKES APPROPRIATE NOTES; CAN ADEQUATELY CONVEY TO OTHERS THE GIST OF WHAT WAS SAID OR DONE |

**WRITTEN**

| **EXCELLENT** | **SATISFACTORY** |
|---|---|
| *READILY* UNDERSTOOD IN SINGLE RAPID READING BY INTENDED AUDIENCE | *ADEQUATELY* UNDERSTOOD IN SINGLE READING BY INTENDED AUDIENCE |
| *READILY* LEGIBLE BOTH IN FORM AND CONTENT; STRONG COMMAND OF WRITTEN ENGLISH | FORM AND CONTENT ARE GENERALLY LEGIBLE, WITH FEW SPELLING OR GRAMMAR ERRORS |
| STYLE VARIED, SIMPLE, TO THE POINT; *USES ACTIVE VOICE WELL* | *REASONABLY* SIMPLE STYLE; GENERALLY USES ACTIVE VOICE |
| BOTTOM LINE UP FRONT; IDEAS CLEAR AND COMPELLING | *USUALLY* HAS BOTTOM LINE UP FRONT—GENERALLY CLEAR |
| STAYS ON TOPIC WELL; CLEAR, CONCISE | *GENERALLY* STAYS ON TOPIC |
| WRITING CONSISTENTLY ACHIEVES STATED PURPOSE; CLEARLY AND COMPLETELY MEETS REQUIREMENTS | WRITING USUALLY ACHIEVES STATED PURPOSE; ADEQUATELY MEETS REQUIREMENTS |

***B. Develops*** – Taking actions to foster team work, encourage initiative, and to accept personal responsibility, while demonstrating care.

## 14) Creates a Positive Environment (CP)

### Creates a positive cultural and ethical environment

**CREATES A POSITIVE ENVIRONMENT (CP)**

| EXCELLENT | SATISFACTORY |
|---|---|
| *EFFECTIVELY* IMPROVES THE COLLECTIVE PERFORMANCE OF THE UNIT IN COMPLIANCE WITH AND SUPPORT OF ORGANIZATIONAL GOALS | *ATTEMPTS* TO IMPROVE THE COLLECTIVE PERFORMANCE OF THE UNIT AND TO COMPLY WITH OR SUPPORT ORGANIZATIONAL GOALS |
| *INSPIRES* BY EXAMPLE; MOTIVATES PEOPLE TO WORK TOGETHER; PROMOTES AND PARTICIPATES IN TEAM ACHIEVEMENT | ACTIVELY WORKS WELL WITH OTHERS; ACTIVE PARTICIPANT IN TEAM EFFORTS; EFFECTIVE TEAM PLAYER |
| FOSTERS ETHICAL CLIMATE; ENSURES PEOPLE ARE TREATED FAIRLY | *ATTEMPTS TO* FOSTER ETHICAL CLIMATE AND TREAT PEOPLE FAIRLY |
| COOPERATIVE, DILIGENT FOLLOWER; HELPS BUILD THE LARGER TEAM (NEST HIGHER UNIT) AS WELL AS OWN TEAM | GIVES SAME GOOD EFFORT WHETHER IN CHARGE OR NOT; CONTRIBUTES TO SUCCESS OF OWN TEAM |
| *HELPS OTHERS* AFTER OR WHILE COMPLETING OWN WORK | GETS OWN SHARE OF WORK DONE |

| | |
|---|---|
| *ADOPTS* UNPOPULAR HIGHER HEADQUARTERS DECISIONS AS OWN; GETS TEAM TO ACT ACCORDINGLY | PROPERLY EXECUTES UNPOPULAR HIGHER HEADQUARTERS DECISION |
| WILLINGLY ACCEPTS, ACTS ON TASKS EVEN ON SHORT NOTICE | ACCEPTS AND ACTS ON ASSIGNED TASKS |
| EFFECTIVE AT WORKING UP THE CHAIN OF COMMAND TO GET PROBLEMS SOLVED THERE | *ATTEMPTS* TO WORK UP THE CHAIN OF COMMAND WHEN A PROBLEM MUST BE SOLVED THERE |
| SPURS THE TEAM TO REMAIN POSITIVE IN CONFUSING OR CHANGING SITUATIONS | REMAINS POSITIVE WHEN SITUATION IS CONFUSED OR CHANGING |

## 15) **Prepares Self (PS)**

### Self-study, self-development and becoming multi-skilled

**PREPARES SELF (PS)**

| **EXCELLENT** | **SATISFACTORY** |
|---|---|
| EXPANDS PERSONAL AND UNIT KNOWLEDGE AND CAPABILITIES | ATTEMPTS TO EXPAND PERSONAL AND UNIT KNOWLEDGE AND CAPABILITIES |
| READILY APPLIES LESSONS LEARNED LEARNED | ATTEMPTS TO FIND WAYS TO APPLY LESSONS |
| CONSISTENTLY LISTENS ACTIVELY; ASKS INCISIVE QUESTIONS | LISTENS; ASKS SOUND QUESTIONS |
| *ACTIVELY SEEKS* FEEDBACK ON PERFORMANCE FROM SUPERIORS, PEERS AND SUBORDINATES; | ACCEPTS FEEDBACK FROM OTHERS; SEEKS TO IMPROVE PERFORMANCE WITH IT |

| | |
|---|---|
| INCORPORATES IT FOR TEAM AND SELF-IMPROVEMENT | |
| SHOWS MATURE LEVEL OF SELF-AWARENESS | EXHIBITS SOME ELEMENTS OF REASONABLE SELF-AWARENESS |

## 16) Develops Others (DO)

### Growing others into competent and confident leaders

**DEVELOPS OTHERS (DO)**

| **EXCELLENT** | **SATISFACTORY** |
|---|---|
| *EFFECTIVELY* EXPANDS AND ENHANCES THE SELF-CONFIDENCE, COMPETENCE AND SELF-ESTEEM OF SUBORDINATES | ATTEMPTS TO EXPAND OR ENHANCE SELF-CONFIDENCE, COMPETENCE OR SELF-ESTEEM OF SUBORDINATES |
| GUIDES & COACHES SUBORDINATES; ENCOURAGES INITIATIVE; LEADS BY EXAMPLE; INSPIRES PEOPLE; DESIGNS AND/OR LEADS EFFECTIVE WAYS TO PRACTICE | FREQUENTLY COACHES SUBORDINATES; OFTEN ENCOURAGES INITIATIVE; VISIBLY TRIES TO LEAD BY EXAMPLE; MAKES SOME EFFORT TO DESIGN WAYS TO PRACTICE |
| *CLEARLY SPECIFIES* STANDARDS AND EXPECTATIONS; USES RECOGNITION OR REWARD TO EFFECTIVELY ENCOURAGE EXCELLENCE | GENERALLY SETS EXPECTATIONS; SHOWS EVIDENCE OF DEALING APPROPRIATELY WITH BOTH GOOD AND POOR PERFORMANCE |
| *ACTIVELY CREATES* OR CONTRIBUTES TO A POSITIVE COMMAND CLIMATE | CONTRIBUTES TO A POSITIVE COMMAND CLIMATE |
| BUILDS ON SUCCESSES, ACTIVELY SEEKS TO IMPROVE UPON WEAKNESSES | *ATTEMPTS* TO BUILD ON SUCCESSES AND IMPROVE UPON WEAKNESSES |

*C. Achieves* – Sets objectives and focuses on mission accomplishment.

## 17) **Gets Results (GR)**

### Developing and executing plans while providing direction, guidance and clear priorities to guide others towards mission accomplishment

**GETS RESULTS (GR)**

| **EXCELLENT** | **SATISFACTORY** |
|---|---|
| *THOROUGHLY* COMPLETES ASSIGNED ASKS TO MEET STANDARD WITHIN COMMANDER'S INTENT, WHILE CARING FOR PEOPLE AND RESOURCES | *GENERALLY* COMPLETES TASKS TO STANDARD AND WITHIN COMMANDER'S INTENT WHILE ADEQUATELY CARING FOR PEOPLE AND RESOURCES |
| *EXCEEDS* SOME STANDARDS | *GENERALLY* MEETS MINIMUM STANDARDS |
| MAKES SMART USE OF AVAILABLE TIME FOR PREPARATIONS, CHECKS AND REHEARSALS | *MAKES SOME* USE OF AVAILABLE TIME FOR PREPARATION, CHECKS AND REHEARSALS |
| MAINTAINS 100% ACCOUNTABILITY OF PEOPLE AND EQUIPMENT; ALWAYS CLEAR WHO IS SUPPOSED TO DO WHAT | *GENERALLY* SHOWS GOOD ACCOUNTABILITY OF PEOPLE AND EQUIPMENT; KEEPS ADEQUATE TRACK OF WHO IS SUPPOSED TO DO WHAT |
| ESTABLISHES AND EFFECTIVELY USES PROCEDURES FOR MONITORING COORDINATING AND REGULATING SUBORDINATES' ACTIONS; MAKES HAPPEN RIGHT THE 1ST TIME | *GENERALLY* ESTABLISHES AND USES PROCEDURES FOR MONITORING, COORDINATING AND REGULATING SUBORDINATES' ACTIONS; HELPS THINGS HAPPEN |

| | |
|---|---|
| OVERCOMES OBSTACLES AND DIFFICULTIES; ENCOURAGES INITIATIVE; THRIVES IN FLUID ENVIRONMENT | *ADEQUATELY* COPES WITH OBSTACLES OR DIFFICULTIES; SHOWS SOME INITIATIVE AND RESILIENCE; HANDLES FLUID ENVIRONMENT |
| ENSURES SUBSTANDARD WORK IS REDONE UNTIL CORRECT | *WILLING* TO MAKE CORRECTIONS (POINTS OUT SUBSTANDARD WORK) |

Now that you have read the differences between satisfactory and excellent performance, go back and look at the italicized words in more detail and observe your peers when they are in charge to see if you can tell the difference. Since everyone's level of understanding is the same, it is necessary to ensure that you become very familiar with what the difference is between these performance levels. If you are unsure, ask your Cadre to point out specific incidences of when a behavior would be satisfactory as opposed to excellent. Therefore, if you can understand the difference, you will be able to provide excellent leadership skills by demonstrating exactly what those "Elusive Evaluators" are looking for in a stellar Cadet.

# 4 Evaluations and the Leader Assessment Program

While at LDAC, you will have a minimum of six Leadership opportunities that include two Garrison positions (one evaluation may be while performing operations in the Forward Operations Base (FOB))that last approximately twenty-four hours in duration, one FLRC that lasts approximately forty-five minutes, two SSTX evaluations that last two hours each and one PSTX leadership position that lasts approximately eight to twelve hours. Cadet Leadership positions range from Squad Leader to Company Commander and are evaluated by several different Evaluators throughout LDAC. The goal is to have six different Cadre assess your leadership potential to eliminate a biased opinion and afford you the full opportunity to succeed even if you have a bad day. Your Evaluators will be members of four different committees that are responsible for certain evaluations. These responsibilities are: Garrison evaluations (CO, XO, 1SG, PL, PSG, SL positions) will be conducted by your Platoon TAC Officer (PTO), Platoon TAC Non-Commissioned Officer (PTNCO) and Platoon TAC Lieutenant (PTLT). FLRC

Leadership Evaluation (SL position) will be evaluated by newly promoted 2nd Lieutenants from across the nation who have volunteered or been assigned to assist at LDAC prior to attending BOLC II. Finally, Cadets receive three field evaluations consisting of two SSTX evaluations (SL position) and one PSTX evaluation (PL, PSG or SL positions) conducted by Evaluators from the Squad STX and Patrolling Committee.

**Leadership Insight 2:** When in charge, **BE IN CHARGE!** How many times have you heard that and asked yourself "What does that mean?" Here are some guidelines for taking charge:

1) Don't let another Cadet make decisions for you.

2) Use subordinate leaders and give them guidance and intent to accomplish the mission.

3) Be proactive and don't wait for tasks to be directed; act in accordance with commander's guidance and intent and refer back to Rule # 2.

4) If there is a lack of commander's guidance, seek guidance; make recommendations so you control the decision, then refer back to Rule # 2.

5) Lead by example; peers and subordinates will emulate what you do; actions are louder than words.

6) Show a genuine concern for your peers when in charge, do what is best for the squad, platoon, company, etc. and don't lead for a grade.

7) Control the situation! Make a decision even if it is the wrong decision and accept responsibility.

Once at LDAC, Cadets must **Understand the Operational Environment**! This is the first of twelve Special Operations Imperatives and is very significant for your success. Understanding the Operational Environment encompasses several inherent tasks such as situational awareness, mission preparation, employing Tactics/Techniques/Procedures (TTPs) and cultural sensitivity. However, the most critical thing to understand in the LDAC operational environment is to know and learn your Evaluator! Each Cadre has different experiences and put different values on certain tasks. Getting an "S" with LDAC credit is fairly easy as long as you demonstrate leadership actions that show potential in terms of what is expected from an MS III Cadet. However, if you are going to get an "E" Rating at LDAC, you must go above and beyond what is expected from an MS III Cadet. You must understand what your Evaluator is looking for to increase your chances of gaining an "E" Rating. As with any job in life, you must manage your bosses by understanding their likes and dislikes or "pet peeves" to set you up for success and keep you from being noticed in a negative way. Below are some actions that will assist you in learning the operational environment but by no means am I am not telling you to be a KISS ASS or suck up to your Evaluators because that will do nothing but piss them off and get you an "N"! You will be able to apply this methodology to Garrison and Field Operations at LDAC by observing other Cadets' evaluations to determine how an Evaluator is grading; for the future Military Intelligence Officers, this is known as Intelligence Preparation of the Battlefield (IPB). If it is during SSTX, you will have the same Evaluator the entire day. You must learn what the lane Evaluator is looking for from the first few Cadets who are in charge. In Garrison you will have more time to evaluate the Cadre than in the field, but either way, you have to do your homework. Here are some insights to making your assessment and "managing your boss":

1) Know your operational environment or situational awareness. Watch other Cadets when they are in a leadership position. Evaluate their leadership and ask yourself if you were in charge, would you be doing the same actions or would you be doing something less or more. After the leadership position is over, pretend that you are the Evaluator and rate the Cadet with an E, S or N; of course, this rating is between you and no one else. Once the Cadet receives an evaluation, ask how they did and what the Cadre said to improve. Don't worry about upsetting someone! Cadets will talk in the platoon and if they don't, then you can almost bet it was an S or N.

2) Observe your Cadre actions. Your Cadre will not give you direct answers but they will give tidbits of information during AARs and during counseling. That is why it is important to talk to Cadets who have been evaluated previously.

3) Listen to what the Cadre says during your Leadership Counseling. Cadre that are evaluating you are almost always observing your actions even when you don't think they are. Al-

though you may think the Cadre has left the area or is not watching, this is usually not the case. Other Cadre observe your actions and give your Evaluator a back brief of your actions while they were away. Additionally, your Evaluator will review your previous Blue Card to see if you have improved or applied recommendations. During each counseling session the Cadre will tell you how to improve and increase your leadership potential so that you can work toward a better evaluation next time. Listen to what is said! During each counseling session, the Cadre point out your weakest Leadership Performance Indicators and what actions will allow you to improve. Make these corrections during your next leadership evaluation and you will see results. Finally, if a Cadre does not give you an example or tell you how you can improve, ask them! If YOU fail to ask them for feedback, then YOU failed to improve your leadership ability.

4) Create a platoon AAR book on previous leadership's lessons learned. Have all previous leadership write notes on what the Evaluator told them to improve and when it is your turn to lead, read the AAR notes in the book before taking charge and then apply them. If your TAC told the previous Cadet to be more assertive by increasing command presence at critical events, then you must identify the critical events and become more involved. For example, if the previous Cadet Leadership were told they must ensure 100% accountability, and the Cadre continues to stress this point, then it is apparent the Cadre feels accountability is an important leadership task. One way to ad-

dress the problem is to create a sign out roster and ensure Squad Leaders know the location of their members. If an Evaluator told the last leadership that a load plan should have been created during the planning phase, ensure you create one during the planning phase of your leadership.

Bottom Line – Lead to lead and not for a grade. If you are looking out for the platoon's best interest, showing a genuine concern for others and being proactive in accomplishing the mission, you will get the "E" you deserve. Don't hesitate, don't allow others to take charge, be aggressive but not overbearing and more importantly, "Let your actions speak for themselves!"

### Evaluation Cards, Spot Reports and Cadet Evaluation Reports (CER)

The Cadet Self Assessment Report known as a Yellow Card offers Cadets the opportunity to make an assessment of their performance during Garrison leadership positions. In my experience, most Cadets do not put a lot of time or effort into these cards because they feel as if they are checking the block and their TAC will not look at the card. I will tell you, "THIS IS NOT THE CASE!" The way Cadets write their Yellow Cards can make a positive or negative impression about how intelligent they are without directly talking to them. Since Yellow Cards are only written in Garrison Operations and Cadets are at multiple training sites with a lot of moving parts, Evaluators may miss a specific event and the reason to why certain behavior occurred. The Yellow Card may clarify why actions were taken resulting in changing the Cadre's perception to what was observed. Another instance is that a Cadre may be Missing in Action (MIA) for some reason or another and the Cadet's entire evaluation will be written from the Yellow Card. Since most Cadets and some Cadre do not understand the power of the Yellow Card, an explanation is warranted. To start, the Yellow Card is divided into three sections and each of these sections has a purpose.

Section I is the *Summary of Performance* and demonstrates good written communication skills, the ability to formulate sentence structure and ensure that the Cadets can summarize tasks, actions and results of their leadership rotations. Additionally, it is the last chance for the Cadets to sell their leadership skills to the Cadre. Although Cadre usually observes 90% of the behavior during your leadership evaluation, they may miss something important that could improve your chances of receiving an "E" instead of an "S" or vice versa. A well-written Yellow Card can increase the communication dimension from an "N" to an "S" or from an "S" to an "E." In the same respect, it can be reversed! This section may also inform the Cadre of an unobserved event that may increase your overall rating. If a Cadre member did not see an event but reads it on the yellow card, they will more than likely begin to ask questions to see how that event occurred and what your actions were during that event. Of course, the Cadre will more than likely not ask you specifically about the event but your peers are a plethora of information and can verify if what your wrote is correct. This affords you the opportunity to ensure that the Cadre is aware of everything that occurred. The best way to formulate section I of the Yellow Card is to follow the STAR, TAR, TAR format.

STAR is an acronym for Situation, Task, Action, and Result.

**Situation** - One to two sentences about what took place during your leadership rotation. Each leadership position has a single situation that encompasses all the major events during the rotation. The situation should address the major events that took place during your leadership evaluation.

**Task** - Major events occurring during the leadership rotation. These tasks support the major events listed in the situation. Addressing two to three tasks per major event should be sufficient.

**Action** - What you did or did not do as a leader to ensure the task was accomplished or not accomplished.

**Result** - The outcome of the leader action.

***Section I Example:***

*S – Platoon Leader during Training Day 5 (D5) consisting of TLPs, land navigation training, barracks maintenance, weapons draw, safety brief, movement to and from the land navigation site, land navigation day testing, land navigation retraining, accountability and AAR.*

*T – Conduct Troop Leading Procedures*

*A – Received Company WO, developed & issued Platoon WO to subordinate leaders giving all available information at the time to include training guidance. Took copious notes and asked pertinent questions while receiving Company OPORD. While adhering to the 1/3 -2/3 rule and the 5 paragraph OPORD format, untilized visual training aids to ensure all Squad Leaders were briefed and understood my intent and guidance.*

*R – The platoon was briefed and informed on specific tasks, prepared for training and ready for lights out by 2200.*

*T – Conduct Land Navigation Training.*

*A – Issued intent to for each squad to conduct land navigation training prior to lights out; ensured PSG developed training plan rotation consisting of plotting, use of compasses, route planning & use of attack points; coordinated with other platoons for training site de-confliction and attended all training.*

*R – Squads attended land navigation training IAW training rotation with no conflict and all squads completed day land navigation testing with a 90% pass rate due to receiving land navigation training before negotiating the course.*

* Continue to use this format for the remainder of the events that took place during the leadership evaluation.

**Leadership Insight 3:** Use the STAR, TAR, TAR format; write accurately and legibly using bullet format or sentence form to ensure that you are short, clear and concise.

*Section II is a Cadet's Strongest and Need to Improve Attributes or Core Leader Competencies,* which is used to communicate the Cadet's perception of his or her strongest and weakest traits to the evaluator. Many Cadets arrive at LDAC with misconceptions about their leadership ability. Cadets may not realize that they are lacking in leadership areas or they may not know the difference between a solid leadership performance and a weak leadership performance. Of course, there are many different scenarios that cause this denial or unawareness but from my experience it is caused by the quality of training presented, varied demographics or the intensity amongst ROTC programs. Because of the variables, Cadets may come from a ROTC Program that is not of the same caliber as others. Cadets who were at the top of their class back at their home campus due to low standards may find themselves below average with other Cadets in their platoon at LDAC. I call this a symptom of being a big fish in a small pond! Another scenario may be that they did not receive an honest assessment of their leadership abilities from their Cadre. Many Cadre in ROTC do not give Cadets an honest assessment because they don't want to be the bad person or they feel the Cadet can't handle the truth. Of course there are a thousand other variables than can make up the Cadet's perception but the bottom line is the Yellow Card allows a Cadre member to see what the Cadet's perception is about his or her leadership ability during a specific leadership evaluation. Cadets must make an Overall Self-Assessment of their leadership performance in this section. This does not automatically indicate that the Cadet will receive this rating. This gives the Evaluator an opportunity to see the Cadets perception.

*Section II Example*

**My Strongest Attributes and /or Core Leadership Competencies:**

*MB – Excellent Command Presence throughout the leadership rotation demonstrated by leading platoon and making on spot corrections.*

*SJ – Ability to make quick decisions at departure time to ensure transportation was not delayed*

**Attributes and /or Core Leadership Competencies I Need to Improve:**

*DK – Unable to perform Company Level Drill & Ceremony Commands with confidence*

*Section III requires the evaluated Cadet to write down and regurgitate what the strongest and need to improve Attributes and /or Core Leadership Competencies (previously identified as LPIs) were addressed by the Evaluator. This provides the Evaluator positive feedback to see if the Cadet understood what was said. To keep Cadets honest, they are not allowed to copy the comments or take the Blue Cards with them from the counseling session. This technique gives the Cadet the best opportunity to improve their leadership ability during the next evaluation so that they can address their deficiencies or reinforce the positive marks.*

*Section III Example*

**Attributes and/or Core Leader Competencies to Improve (As Noted by Evaluator):**

*DK/CF – Was unable to execute Company D&C Commands with confidence to Improve (as noted by Evaluator):*

***Actions I plan to take:*** *I plan to watch other Cadets in leadership positions to improve my overall D&C knowledge to include practicing in a mirror to reduce nervousness. Improving my organizational skills by tasking subordinate leaders will allow me to reduce stressful situations and I will seek out additional help from Cadet "At Ease" to improve my ability in controlling my emotions in stressful situations.*

## The Yellow Card

**CADET SELF-ASSESSMENT REPORT**
(ROTC Cdt Cmd Reg 145-3)

REQUIREMENTS CONTROL SYMBOL
ATCC-122

| Cadet | Unit | Unit Position | Length (hours) | Date |
|---|---|---|---|---|
| DRAGGER, KNUCKLE | 1/A/1/1 | PL | 24 | ddmmyy |

SUMMARY OF PERFORMANCE (Describe the Situation, assigned Task, the Action taken, and the Results).

**S**: *Platoon Leader during D5 consisting of TLPs, land navigation training, barracks maintenance, weapons draw, safety brief safety brief, movement to and from the land navigation site, land navigation day testing, land navigation retraining, accountability and AAR.*

**T**: *Conduct Troop Leading Procedures*

**A**: *Received Company WO, developed & issued Platoon WO to subordinate leaders giving all available information at the time to include training guidance. Took copious notes and asked pertinent questions while receiving the Company OPORD. While adhering to the 1/3 - 2/3 rule and the 5 paragraph OPORD format, utilized visual training aids to ensure all Squad Leaders were briefed and understood my intent and guidance.*

**R**: *The platoon was briefed and informed on specific tasks , prepared for training and ready for lights out by 2200.*

**T**: *Conduct Land Navigation Training.*

**A**: *Issued intent to for each squad to conduct land navigation training prior to lights out; ensured PSG developed training plan rotation consisting of plotting, use of compasses, route planning & use of attack points; coordinated with other platoons for training site de-confliction and attended all training.*

**R**: *Squads attended land navigation training IAW training rotation with no conflict and all squads completed day land navigation testing with a 90% pass rate due to receiving land navigation training before negotiating the course.*

CDT CMD Form 156-2—ALL PREVIOUS VERSIONS

**Figure 2 Cadet Self-Assessment Report (Yellow Card)**

My Strongest Attributes and/or Core Leader Competencies: (list 1,2, or 3)

***MB*** *– Excellent Command Presence throughout the leadership rotation demonstrated by leading platoon and making on spot corrections*
***SJ*** *– Ability to make quick decisions at departure time to ensure transportation was not delayed*

Attributes and/or Core Leader Competencies I Need to Improve: (list 1,2, or 3)

***DK*** *– Unable to perform Company Level Drill & Ceremony Commands with confidence*

Overall Self-Assessment (E) S N

Complete self-assessment to this point and turn it in prior to counseling

## Cadet Summary of Counseling

My Strongest Attributes and/or Core Leader Competencies (As Noted by Evaluator):

***MB*** *– Demonstrated Great Command Presence by taking charge without any reservations; good Command Voice*
***RS*** *– Demonstrated the ability to recover after moving company to the wrong loading area by adapting the plan to load vehicles on time*

Attributes and/or Core Leader Competencies to Improve (As Noted by Evaluator):

***DK/CF*** *– Was unable to execute Company D&C Commands with confidence*

Actions I Plan to Take: (Sentences or bullets)
*I plan to watch other Cadets in leadership positions to improve my overall D&C knowledge to include practicing in a mirror to reduce nervousness. Improving my organizational skills by tasking subordinate leaders will allow me to reduce stressful situations. I will seek out additional help from Cadet At Ease to improve my ability in controlling my emotions in stressful situations.*

| Cadet's Signature Knuckle Dragger | Date ddmmyy |
|---|---|

**Figure 3 Cadet Self-Assessment Report - Reverse Side (Yellow Card)**

## The Blue/Green Card

### LEADERSHIP ASSESSMENT REPORT

CADET COMMAND REG 145-3
REQUIREMENTS CONTROL SYMBOL ATCC-122

**PART I – Attributes (what a leader is):** Characteristics that are an inherent part of an individual's total core., physical, and intellectual aspects. Attributes shape how one behaves

**1. Character:** A person's moral and ethical qualities which give a leader motivation to do what is appropriate regardless of circumstances or consequences

| ARMY VALUES (Comments mandatory in Part III for all "NO" entries) | Yes | No | | Yes | No |
|---|---|---|---|---|---|
| **1. LOYALTY:** Bears true faith and allegiance to the U.S. Constitution, the Army, the Unit and other Soldiers | | | | | |
| **2. DUTY:** Fulfills professional, legal, and moral obligation | | | **5. HONOR:** Adheres to the Army's publicly declared code of values | | |
| **3. RESPECT:** Treats others as the should be treated, promotes dignity, consideration, & fairness | | | **6. INTEGRITY:** Does what is right both legally and morally, honest in word and deed | | |
| **4. SELFLESS-SERVICE:** Places welfare of others and Army priorities before self | | | **7. PERSONAL COURAGE:** Faces fear, danger, or adversity | | |
| **EMPATHY:** The ability to see something from another person's point of view, to identify with and enter into another person's feelings and emotions | | | | | |
| **WARRIOR ETHOS:** I will always place the mission first, I will never accept defeat, I will never quit, I will never leave a fallen comrade | | | | | |

Mark "E", "S", or "N" for each dimension. IMPROVE comments in Part III are mandatory when rating of "N" is indicated

**2. Presence**
The impression that a leader makes on others, which contributes to their success in leading them, the image that a leader projects; how others perceive a leader (outward appearance, demeanor, words and actions)

| Code | Dimension | E | S | N |
|---|---|---|---|---|
| MB | **Military Bearing** — Projecting a commanding presence and professional image of authority | X | S | N |
| PF | **Physically Fit** — Having sound health, strength, and endurance that supports one's emotional health and conceptual abilities under stress | X | S | N |
| CF | **Confident** — Projects self-confidence and certainty, demonstrates composure and poise, calm and collected, possesses self control of emotions | E | X | N |
| RS | **Resilient** — Showing a tendency to recover quickly from setbacks, shock, adversity, stress or injury while maintaining a mission and organizational focus | E | X | N |

**3. Intellectual Capacity**
The ability to draw on the mental tendencies and resources that shape a leader's conceptual abilities and impact of effectiveness, which then are applied to one's duties and responsibilities.

| Code | Dimension | E | S | N |
|---|---|---|---|---|
| MA | **Mental Agility** — Flexibility of mind, a tendency to anticipate or adapt to ever changing conditions, improvisation | X | S | N |
| SJ | **Sound Judgment** — Assesses situations and draws feasible conclusions makes sound and timely decision | X | S | N |
| IN | **Innovation** — Ability to introduce something new, is original in thoughts and ideas, creative | X | S | N |
| IP | **Interpersonal Tact** — Effectively interacts with others, possesses the capacity to understand personal interactions with others, awareness of how others see you | E | S | N |
| DK | **Domain Knowledge** — Possessing facts, beliefs, and logical assumptions in relevant areas, technical, tactical cultural and geopolitical knowledge | E | X | N |

**PART II – Core Leader Competencies (what a leader does):** Works to lead others, develops themselves, their subordinates and organizes to achieve mission accomplishment

**1. Leads**
The application of character, presence, intellect and abilities while guiding others toward a common goal and mission accomplishment.

| Code | Dimension | E | S | N |
|---|---|---|---|---|
| LD | **Leads Others** — Motivates, inspires, and influences others to take initiative, to work toward a common purpose, to accomplish critical tasks and to achieve unit objectives | X | S | N |
| EI | **Extends Influence beyond CoC** — Uses indirect means to influence others outside normal chain of command involves diplomacy, negotiation, conflict resolution and mediation | E | S | N |
| LE | **Leads by Example** — Provides the example to others, serves as a role model maintains high standards in all aspects of behavior and character | E | X | N |
| CO | **Communicates** — Clearly expresses ideas to ensure understanding actively listens to others, and practices effective communication techniques | X | S | N |

**2. Develops** Taking actions to foster team work, encourage initiative, and to accept personal responsibility, while demonstrating care

| Code | Dimension | E | S | N |
|---|---|---|---|---|
| CP | **Creates a Positive Environment** — creates a positive cultural and ethical environment | E | S | N |
| PS | **Prepares Self** — Self-study, self-development and becoming multi-skilled, ensues they are prepared to lead | E | S | N |
| DO | **Develops Others** — Encourages and supports others to grow as individuals & teas, prepares others for success, makes the organization more versatile | E | S | N |

**3. Achieves**
Sets objectives and focuses on mission accomplishment

| Code | Dimension | E | S | N |
|---|---|---|---|---|
| GR | **Gets Results** — Structure what needs to be done so results are consistently produced, developing and executing plans while providing direction, guidance and clear priorities towards mission accomplishment, manages the resources required for mission accomplishmentd | X | S | N |

**Figure 4. Leadership Assessment Report Part I and II (Blue/Green Card)**

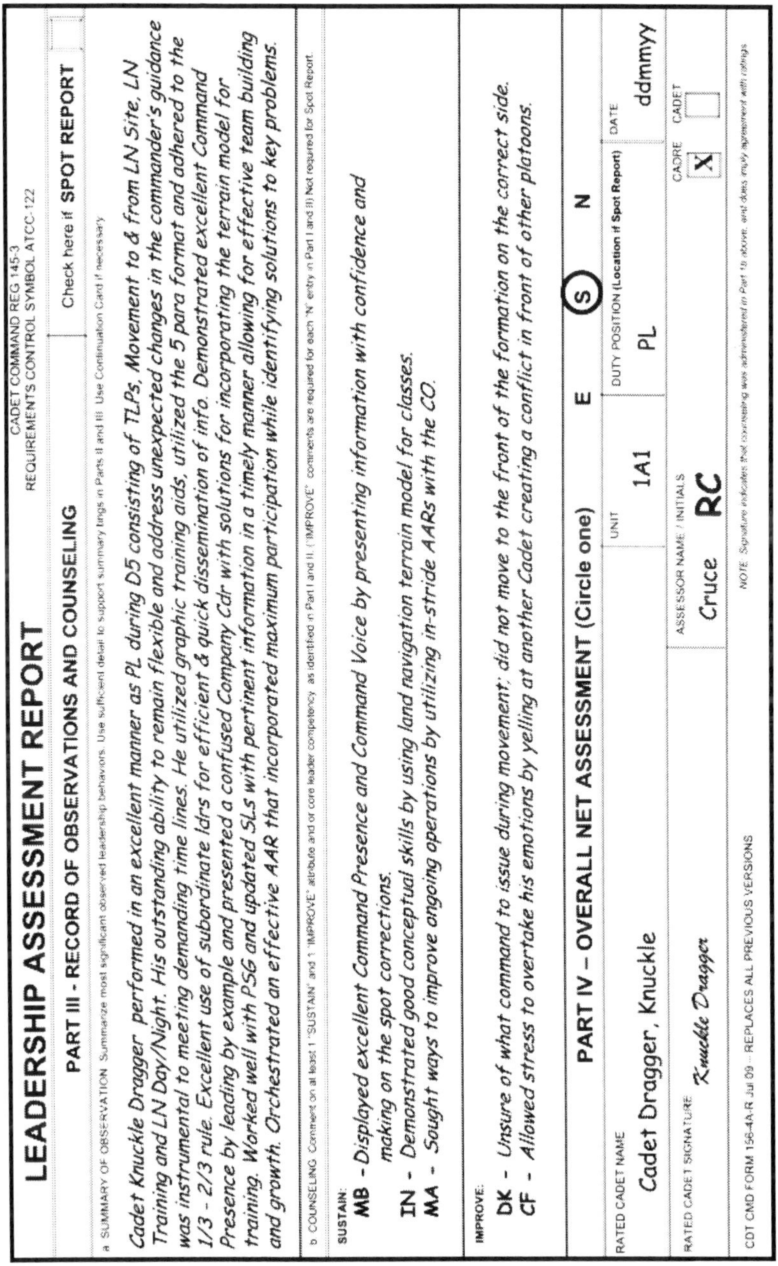

# LEADERSHIP ASSESSMENT REPORT

CADET COMMAND REG 145-3
REQUIREMENTS CONTROL SYMBOL ATCC-122

PART III - RECORD OF OBSERVATIONS AND COUNSELING

Check here if SPOT REPORT

a. SUMMARY OF OBSERVATION. Summarize most significant observed leadership behaviors. Use sufficient detail to support summary tings in Parts II and III. Use Continuation Card if necessary.

*Cadet Knuckle Dragger performed in an excellent manner as PL during D5 consisting of TLPs, Movement to & from LN Site, LN Training and LN Day/Night. His outstanding ability to remain flexible and address unexpected changes in the commander's guidance was instrumental to meeting demanding time lines. He utilized graphic training aids, utilized the 5 para format and adhered to the 1/3 - 2/3 rule. Excellent use of subordinate ldrs for efficient & quick dissemination of info. Demonstrated excellent Command Presence by leading by example and presented a confused Company Cdr with solutions for incorporating the terrain model for training. Worked well with PSG and updated SLs with pertinent information in a timely manner allowing for effective team building and growth. Orchestrated an effective AAR that incorporated maximum participation while identifying solutions to key problems.*

b. COUNSELING. Comment on at least 1 "SUSTAIN" and 1 "IMPROVE" attribute and or core leader competency as identified in Part I and II. ("IMPROVE" comments are required for each "N" entry in Part I and II) Not required for Spot Report.

SUSTAIN:

*MB - Displayed excellent Command Presence and Command Voice by presenting information with confidence and making on the spot corrections.*

*IN - Demonstrated good conceptual skills by using land navigation terrain model for classes.*

*MA - Sought ways to improve ongoing operations by utilizing in-stride AARs with the CO.*

IMPROVE:

*DK - Unsure of what command to issue during movement; did not move to the front of the formation on the correct side.*

*CF - Allowed stress to overtake his emotions by yelling at another Cadet creating a conflict in front of other platoons.*

PART IV – OVERALL NET ASSESSMENT (Circle one) E (S) N

| RATED CADET NAME | UNIT | DUTY POSITION (Location if Spot Report) | DATE |
|---|---|---|---|
| Cadet Dragger, Knuckle | 1A1 | PL | ddmmyy |

| RATED CADET SIGNATURE | ASSESSOR NAME / INITIALS | CADRE | CADET |
|---|---|---|---|
| *Knuckle Dragger* | Cruce RC | X | |

CDT CMD FORM 156-4A-R Jul 09 – REPLACES ALL PREVIOUS VERSIONS

NOTE: Signature indicates that counseling was administered in Part 1b above, and does imply agreement with ratings.

**Figure 5: Leadership Assessment Report - Reverse Side (Blue/Green Card)**

The Yellow Card is an important step in the Garrison Evaluation Process by allowing Cadets the opportunity to conduct self assessment before receiving The Leadership Assessment Report, known as the Blue/Green Card. The Blue/Green card is used by Evaluators to write a synopsis of the Cadet's Leadership Behavior for each of the six designated leadership evaluations. A Cadet's behavior is the primary driver in writing evaluations and includes watching how the Cadet addresses different situations. Since Cadets have different skill sets and different experience levels, their behavior and actions allow for an honest assessment. Due to the vast difference in Cadets' experiences, LDAC looks at the Cadets' behavior instead of their ability to accomplish a mission with perfection. Therefore, in reality, what is really being evaluated is the Cadet's potential to become a leader. The only difference between a Blue and Green Card is their color. The Blue Card is used during Garrison evaluations and the Green Card is used for field evaluations (FLRC, SSTX & PSTX evaluations). Both cards format is exact and divided into three parts:

*Part I (Attributes (What a Leader Is))* - These attributes encompass the seven Army Values, Empathy, Warrior Ethos and nine leadership behaviors under the Presence and Intellectual Capacity areas of the Army Leadership Requirements Model. These leadership behaviors should be supported by the observations noted in Part III.

*Part II (Core Leader Competencies (What a Leader Does))* - These core leadership competencies encompass the remainder eight leadership behavioral behaviors that fall under the areas of Leads, Develops and Achieves.

*Part III is Records of Observations and Counseling* which allows evaluators to write a summary of the Cadets' behavioral observations and counsel the Cadets on those leadership behaviors. The counseling should address at least one sustain and one improve attribute and/or core leader competency in Part I and II. Improve comments are mandatory for any "N" entry in Part I or II.

*Part IV is the Overall Net Assessment of the Cadet's performance (E, S, or N). A Cadet attending LDAC can receive one of three ratings in each of the seventeen leadership behaviors within the areas of Presence, Intellectual Capacity, Leads, Develops or Achieves. These ratings are based on what a MS III Cadet should know after three years of training at ROTC. This is very important to keep in mind because this is how the Excellence "E", Satisfactory "S" and Needs Improvement "N" assessments are determined. Although each Cadre Evaluator will have different experiences, they all are calibrated on what skills MS III Cadets should possess while attending LDAC. From a Cadet's perspective, this is very arbitrary!*

## COLD HARD FACT 4:

A Cadet can receive an "N" in Leadership Performance Indicator (LPI) on a Blue/Green Card and still receive an overall "E."

## Ratings Explained

The *Expected Excellence "E" Rating* is the best rating a Cadet can get for an LPI or an Overall Assessment at LDAC. A Cadet must receive three out of six Overall Net Assessments of Excellence "E" Ratings during leadership evaluations and a minimum of 50% "E" marks in the Seventeen Leadership Performance Indicators (LPI) to receive an Overall Excellence Rating at LDAC. Some situations that may keep you from getting an "E" with the required three Overall "E" Net Assessments are failing the APFT, Land Navigation day/night/written, low peer rankings or an overall N for a Leadership Assessment.

The *Sad but Satisfactory "S"* Rating is a great rating and should be renamed to Successful in my opinion. A MS III Cadet that attends LDAC and walks away with an Overall Net Assessment of "S" is successful and can be a great leader in the Army. The Cadet may not rank as high on the National Order of Merit List (OML) but he or she will gain LDAC credit and be ready for commissioning after graduating from college. The "S" rating is earned when a Cadet has completed all the required events and received less than three Overall Net Assessments of an "E" or less than 50% "E" marks in the Seventeen Leadership Indicators (LPI).

The *Devastating Needs Improvement "N" Rating* is exactly what it says and nothing more! There are two outcomes for Cadets who receive an "N" at LDAC. The first outcome is a Cadet receives an Overall "N" Rating with LDAC credit. If this happens, the Cadet has met the minimum standard to receive LDAC credit but the Cadet has a long road to walk prior to commissioning as a 2LT. The Cadet must bust his or her ass to show the PMS that he or she is worthy and has the potential to become an Army Officer. One key note to remember is, the PMS has the right and authority to disenroll a Cadet from ROTC even if he or she receives LDAC credit! Therefore, a Cadet with an "N" Rating and LDAC credit must show improvements the following year before he or she is commissioned. Finally, receiving an "N" does not mean the Cadet is a NOGO or that the Cadet will not be a good leader. Instead, the "N" Rating has exposed an individual's weakness which requires a training plan to improve the weakness during the MS IV year.

The second outcome is when a Cadet does not receive LDAC credit. This outcome has two different scenarios and they both have a certain level of suck. The first scenario allows the Cadet to remain in the ROTC program and return to LDAC the following year. If this scenario occurs, it is not the end of the world! Although you lost the battle to the enemies discussed earlier in the book you will retrain, lick your wounds and return for another battle again next year. A Cadet who departs Warrior Forge without LDAC credit was not properly prepared physically or mentally prior to arriving LDAC, received an honor violation or absolutely did not have the will to succeed. The second scenario when the Cadet is disenrolled from the ROTC Program upon return to home campus. If the second scenario occurs, then do some soul searching and ask yourself why your PMS did not want to keep you in the program? Did you give your best or did you only do the minimum requirements for the program? There is a major difference! Afterwards, develop a new course of action and use it as a lesson in life and don't allow yourself to be put in that situation again.

### COLD HARD FACT 5:

Cadets who do not receive LDAC credit can still be commissioned if they return the following year and pass; however, I would not count on it!

*Tracking your Progress* is a very necessary task at LDAC so that you maintain situational awareness of your performance. Evaluators use a Job Performance Summary Card (JPSC) to track each Cadet's performance because there are just too many Cadets to try and remember each and every evaluation. Some Cadets have no idea how they are doing at LDAC because they only remember the overall rating and don't understand that there is another crucial part of the evaluation called Leadership Performance Indicators (LPIs). In my opinion, the Es from LPIs play a larger role in your placement on the OML than the overall LDAC rating. The bottom line is that Cadets must track their progress to know where they stand in performance. A recommendation prior to receiving leadership counseling is to make a diagram in your notebook to track the Seventeen LPIs from each of the six leadership positions. Ensure the notebook is open and write down each LPI rating during each leadership evaluation. Don't be afraid to ask to see the Blue Card or have the Evaluator repeat the marks. By documenting the LPIs and what is said during counseling, you demonstrate that you have a positive attitude and that you are interested in your professional growth. Additionally, by tracking these LPIs you see what leadership behaviors you are weak at or what leadership behaviors you are not addressing to help you prepare for the next evaluation. By tracking your progress, it will be clear what leadership behaviors the Evaluators have not seen from your actions and it will show you what area you need to address during your next leadership position or positive spot report to increase your overall rating. Some Cadets try and place themselves in the right location to stand out in front of Cadre to receive a positive spot reports, a.k.a. "Spot Light Ranger," however, the best way to get a positive spot report is to do the right thing all the time in the absence of Cadre. If you are genuine in your actions and do what is best for the unit (your platoon), the positive spot reports will come with ease. Think of LDAC as a real unit and do what is necessary for the unit to succeed. Nonetheless, you must be genuine in your actions or Cadre and your peers will see right through you! Either way, tracking your progress allows you to know the number of leadership behavioral Es and what Overall Assessment you should receive with certainty.

## Table 2: Cadet LPI Tracker

| Position | MB | PF | CF | RS | MA | SJ | IN | IP | DK | LD | EI | LE | CO | CP | PS | DO | GR | NET |
|---|---|---|---|---|---|---|---|---|---|---|---|---|---|---|---|---|---|---|
| GAR 1 | | | | | | | | | | | | | | | | | | |
| GAR 2 | | | | | | | | | | | | | | | | | | |
| FLRC | | | | | | | | | | | | | | | | | | |
| SSTX | | | | | | | | | | | | | | | | | | |
| PSTX | | | | | | | | | | | | | | | | | | |
| Spt Rpt | | | | | | | | | | | | | | | | | | |
| Spt Rpt | | | | | | | | | | | | | | | | | | |
| TOTAL | | | | | | | | | | | | | | | | | | |

## Table 3: JPSC without Spot Reports

| | PRESENCE | | | | INTELLECTUAL CAPACITY | | | | | LEADS | | | | DEVELOPS | | | ACHIEVES | |
|---|---|---|---|---|---|---|---|---|---|---|---|---|---|---|---|---|---|---|
| Position | MB | PF | CF | RS | MA | SJ | IN | IP | DK | LD | EL | LE | CO | CP | PS | DO | GR | NET |
| GAR 1 | E | E | S | S | E | S | | N | S | E | E | S | S | N | S | | E | E |
| GAR 2 | S | S | E | E | E | E | | | E | S | S | S | S | S | S | S | S | S |
| FLRC | S | S | S | S | S | E | | E | S | E | E | E | E | S | S | S | | S |
| SSTX 1 | E | | E | S | S | S | E | S | S | E | S | E | S | | | S | E | S |
| SSTX 2 | E | S | S | E | S | E | E | S | E | S | S | S | E | | | | S | E |
| PSTX | E | S | E | E | S | E | E | S | S | E | S | S | E | | E | E | | E |
| TOTAL | E | S | E | E | S | E | E | S | S | E | S | S | E | S | S | S | E | S |

*SPOT Reports* can upgrade or degrade a Cadet's Leadership Performance Indicators (LPIs) and overall assessment. The spot reports allow for a Cadre to document good or bad behavior and acts. They are used to evaluate Cadets while not in a leadership role and crucial in the identification of "Team Players," "Spot Light Rangers," and "Blue Falcons." These spot reports do not count as an overall evaluation but they are used in determining the rating of the Seventeen LPIs (See Table 3). This becomes important for Cadets because if they receive three overall "E" leadership evaluations but earn less than 50% Es in the Leadership Performance Indicators (LPI), they receive an overall Satisfactory "S" Rating at camp unless a waiver is submitted to upgrade the overall rating to an "E." Of course this waiver has to be approved through the chain of command; the waiver is submitted by the Platoon TAC Officer (PTO) to the Company TAC Officer (CTO) to the Regimental TAC Officer (RTO) to the Commandant of the Cadets (COC) who is the approval authority. By the way, good luck with that! Over the last couple of years, Cadre were very skeptical of submitting an upgrade. For example, I knew a Cadet that was rated #3 in his platoon, received 10 leadership behavioral Es, maxed the APFT and land navigation but only received 2 Overall "E"s in his leadership positions. The Cadre in that platoon did not submit a migration packet and the Cadet left LDAC with an overall "S" Rating. In my opinion, this was a prime example of Cadre that were either uneducated on the purpose of the migration packet or just lazy. Below shows how a migration packet would flow for approval.

**PTO → CTO → RTO → COC**

## Table 4: JPSC With Spot Reports

| | PRESENCE | | | | INTELLECTUAL CAPACITY | | | | | LEADS | | | | DEVELOPS | | | ACHIEVES | |
|---|---|---|---|---|---|---|---|---|---|---|---|---|---|---|---|---|---|---|
| Position | MB | PF | CF | RS | MA | SJ | IN | IP | DK | LD | EI | LE | CO | CP | PS | DO | GR | NET |
| GAR 1 | E | E | S | S | E | S | | N | S | E | E | S | S | N | S | | E | E |
| SPOT RPT | | | | E | | | | E | | | | | E | | | ***E*** | | |
| GAR 2 | S | S | E | E | E | E | | | E | S | S | S | S | S | S | S | S | S |
| FLRC | S | S | S | S | S | E | | E | S | E | E | E | E | S | S | S | | S |
| SPOT RPT | | E | E | E | | | | E | | | | | | | | ***E*** | | |
| SSTX 1 | E | | E | S | S | S | E | S | S | E | S | E | S | | | S | E | S |
| SPOT RPT | | E | | | E | | | | | | | | | | E | ***E*** | | |
| SSTX 2 | E | S | S | E | S | E | E | S | E | S | S | S | E | | | | S | E |
| PSTX | E | S | E | E | S | E | E | S | S | E | S | S | E | | E | E | | E |
| TOTAL | E | S | E | E | S | E | E | S | S | E | S | S | E | S | S | ***E*** | E | S |

Table 3 is a portion of a Job Performance Summary Card (JPSC) that Cadre use to track the progress of each Cadet. In the above JPSC, the Cadet received three overall "E" and three overall "S" ratings from the six leadership positions and no spot reports. After tallying up the results of the six leadership positions, the Cadet finished LDAC with three Net "E" ratings and eight E marks in the seventeen LPIs, which is less than the 50% required.

Table 4 shows a new result when the Cadet receives three positive spot reports added to the JPSC for giving classes, calling cadence and being a team player by carrying an injured soldier's rucksack during a foot march. One can see the difference because the leadership behavior Developing Others (DO) changes to an "E" which changes the total Leadership Performance Indicator Es from eight to nine, which is more than 50%. The additions of these three Spot Reports with an increase in E marks for a specific LPI, result in the Cadet receiving an overall "E" Rating at LDAC which can make a difference in placement on the National Order of Merit List (OML).

**The Cadet Evaluation Report (CER)** is used to develop and report Cadets' behavior during their MS III year prior to and after attending LDAC. The CER is similar to an Officer Evaluation Report (OER) because it takes into account the seven Army Values and seventeen Leadership Performance Indicators and is used to familiarize Cadets with the Army Evaluation System. The On Campus CER is received prior to attending LDAC and is presented to the Cadet by their Professor of Military Science (PMS). The PMS establishes an Order of Merit List (OML) that takes your MS III year into account by measuring leadership, ROTC participation, ROTC training results and peer rankings to get an overall assessment of your potential to serve in the United States Army and uses this OML to assist in developing the On Campus CER. The On Campus CER focuses on the 17 Leadership Performance Indicators, strengths and weaknesses and gives the Cadet a good assessment before entering the BIG POND of LDAC! The LDAC CER follows the same format as the On Campus CER but is focused on your performance and potential while at LDAC. Although the duration may be shorter, more situations allow behavior to be evaluated on multiple events which usually provides the same assessment as before leaving your campus. In my experience, the results from the On Campus CER and the CER at LDAC resemble each other unless a Cadet has a EUREKA moment between the spring semester and camp. Just like the old saying "a zebra does not change its stripes over night" the same holds true to a Cadet's personality and actions. We referred to this earlier under the section "The Inner Villain" (TIV). When a Cadet takes the information from the two CERs and applies a positive attitude to actively seek out improvement, the CER becomes a powerful tool and develops the Cadet into a much more powerful leader.

**Peer Evaluations** are conducted three times at LDAC; twice at the squad level and once at the platoon level. A squad peer evaluation is conducted approximately around D9 and on D19 and a platoon peer evaluation is conducted the last week. Squad peer evaluations are very valuable to Cadets if they keep an open mind and adjust their behavior from the information received. The squad peer evaluation is internal to the squad and requires each member to rate each Cadet in the squad except for themselves. The peer evaluation requires that three positives and three negatives be written about each of those Cadets so that they have specific areas identified. If you are rated twelve out of twelve, you have an issue. If you are rated higher than six, you have some issues that must be improved. If this is the case, read the positives and negatives that each Cadet stated and develop a course of action to correct the negative. Use the positives to your advantage and ask for constructive criticism from your peers. Showing a genuine interest in self-improvement goes a long way with your squad. The most popular negative comments on the peer evaluations from my experience are: Cadet is not a team player, Cadet lacks command presence, Cadet lacks interpersonal skills, Cadet is immature by joking too much and the Cadet is very indecisive. The most popular positive comments are: Cadet is a team player, Cadet displays confidence, Cadet shows technical and tactical knowledge, Cadet is an active listener and the Cadet remains positive. The platoon peer evaluation does not mandate positive and negatives comments be made but instead it works off of a number system. This becomes very difficult to measure accurately because the personnel in the squads are much closer than outside the squads. Every Cadet ranks every other Cadet in the platoon. Therefore, if you are a silent individual and have not attempted to assist everyone in the platoon or step up and be noticed by providing leadership, it is very likely that you will be peered low. The platoon peer rating is a complex ranking system that uses a mathematical equation but the accuracy is spot on.

**PEER EVALUATION REPORT**
(ROTC Cdt Cmd Reg 145-3)

REQUIREMENTS CONTROL SYMBOL ATCC-122

| CADET (RATER) | UNIT | DATE |
|---|---|---|
| CADET (RATED) | RANKING ______ OF ______ | |

WHAT ARE THIS INDIVIDUAL'S STRONGEST (ONE, TWO, THREE) LEADERSHIP ATTRIBUTES/CORE LEADER COMPETENCIES AND WHY?

WHAT ARE THIS INDIVIDUAL'S WEAKEST (ONE, TWO, THREE) LEADERSHIP ATTRIBUTES/CORE LEADER COMPETENCIES AND WHY?

**Figure 6: CDT CMD Form 156-17-R July 09**

## "Marginal Performer" and "At Risk" Cadets

If a Cadet does not meet the minimum standards during an event, he or she will be identified as a "Marginal Performer." Marginal performance is viewed as not meeting the Warrior Forge minimum standards or a consistent low performance in multiple events. The earlier a Cadet is identified as a Marginal Performer, the faster a plan of action can be determined to increase the Cadet's skills to a minimum standard. If Cadets fall into the marginal performance category, they receive a counseling statement notifying them that they did not meet a minimum standard. They are then placed on Marginal Performance Status and are counseled on a plan of action to correct the deficiency. The Cadet's PTO will initiate the marginal performance counseling statement to include a waiver recommendation to either continue training at LDAC with the opportunity of a retest or be removed from LDAC so that the Cadet can return to their home campus and receiving additional training to better prepare for next year. Immediately following the PTO's counseling session and recommendation, the CTO will interview the Cadet and make a recommendation. Finally, the RTO will interview the Cadet and make a recommendation to the Commandant of Cadets (COC) who will determine if the Cadet remains at Warrior Forge, recycled or returned to the Cadet's home campus for further training.

"At Risk" Cadets are identified as Cadets who have the potential to require multiple waivers or fail to meet Warrior Forge completion criteria. At Risk Criteria may include but not limited to a consistent pattern of low scores and performance (e.g. APFT and Land Navigation Scores), receive an overall assessment of "N" in a leadership evaluation or a projected summary rating of Needs Improvement in a Leadership Performance Indicators or Army Values.

**Migration packets** are not a certainty but do exist if the situation warrants a Cadet to receive a higher Overall Evaluation Grade. Although these situations are very rare, I certainly would not want to rely on this procedure. The bottom line is never to give up and always do your best because miracles do happen every day. Below are some examples of when a migration packet may be initiated:

*Cadet Couch Potato received 3 "E" Ratings on their Overall Net Assessments (ONA) and 12 Es in Leadership Performance Indicators (LPI) but failed the APFT (Automatic N) with a run score of 58 points and total APFT Score of 258 points; Cadet Couch Potato took the second APFT and scored 262 points maxing the Push-ups and Sit-ups and passing the run with a score of 62 points; Result is "N" (Needs Improvement) with LDAC credit with no migration packet.*

*Cadet Almost There received 3 "E" Ratings on their ONA and Less than 50% Es in leadership Performance Indicators; Result is "S" (Satisfactory) and will more than likely not warrant a migration packet.*

*Cadet Lost at Night received 3 "E" Ratings on their ONA and 10 Es in LPIs, passed day land navigation with 8 out of 8, scored 20 out of 20 on written land navigation but failed night land navigation on the first attempt. Cadet Lost at Night passed night land navigation on the second attempt by finding 5 out of 5 points; Result is an "N" (Needs Improvement); may warrant a migration packet.*

*Cadet Bad Day received 4 "E" Ratings on their ONA and 12 Es (more than 50%) in LPIs, passed land navigation written, day and night with 100% but failed the APFT run by 1 second; the Cadet maxed the Push-ups and Sit-ups but felt very ill and was throwing up before the APFT. Cadet Bad Day took the second APFT and scored 290 points maxing the Push-ups and Sit-ups; Result is "N" (Needs Improvement); may warrant a migration packet.*

As a PTO, my recommendation would be to initiate a migration packet on Cadet BAD Day from an "N" to an "E". This migration is based on the fact that the poor performance was due to the illness during the first APFT and is supported by the fact that the Cadet performed outstanding in all other events.

**Leadership Insight 4:** Dedicate time to studying ROTC information by performing daily physical fitness and keeping a positive attitude. Continue to strive for excellence even if you fail an event. You may be able to recover if you keep a good attitude and apply yourself.

Each platoon PTO/PTNCO will select the **Platoon Top 5 Cadets** which adds an additional 1 point to the Cadet's placement on the National Order of Merit List. Each PTO/PTNCO will have different criteria for selecting the top 5 Cadets but for the most part it will be evaluated on overall performance at LDAC. The major events I used to determine the Platoon Top 5 were, APFT, Land Navigation, BRM, Peer Evaluations, Leadership Positions and most importantly what contributions the Cadet gave to the platoon. In every organization, there are those leaders who stand out as team players, always do what is best for the unit, take charge and are sought out by their peers for advice. Cadets are selected as the top 5 in the platoon because they have mastered individual skills, they treat LDAC as a real world scenario and they put 100% effort toward assisting the platoon to succeed even if they do not get the recognition.

**Leadership Insight 5:** Always strive to do your best, develop the less skilled Cadets in the platoon, be a team player and be a leader. But most of all don't have THS (Tiny Heart Syndrome).

**The Coveted RECONDO Badge** is worth an extra ½ point on the National OML and in theory is fairly easy to obtain whether the Cadet is male or female and especially for those with Type A personalities. It also is a great goal to have and can assist in boosting Cadets' overall position on the National OML to assist in getting their branch of choice. The criteria to receive the Recondo Badge are to execute designated events outlined in Cadet Command Circulative (CCR) 145-05 with a certain level of confidence or with a certain score. CCR 145-05, Section 1, Paragraph 6 list the following criteria to receive the RECONDO Badge:

Execute specified Confidence and CWST events presented to standard **on the day of scheduled execution.** Failure to meet the standard for ANY event denotes a "NO GO" for badge qualification.

Achieve an APFT score of 270 or above without retesting, 90 points minimum in each event.

Without retest, achieve a score of 80% or higher on each event (written, day, night) of the Land Navigation proficiency test (16/20 Written, 6/8 stakes Day, 4/5 stakes Night).

Achieve a "GO" on BRM (23 of 40 hits) on their **first or second** attempt.

Complete the CPT Michael Tarlavsky Hand Grenade Assault Course and the 1LT Naioa Hoe Combat Assault Course.

Complete the following 1st Aid tasks to specified standard: CPR, Evaluate a Casualty, Manage the Airway, Control Bleeding (less burns).

Achieve satisfactory or higher performance ratings on BOTH Squad STX evaluations.

Successfully complete WF without a performance waiver. Medical waivers are acceptable.

Receive satisfactory or higher summary ratings for all 17 Leadership Performance Indicators and Attributes, as reported on the Cadet Command Form 67-9, Cadet Evaluation Report.

Meet Height/Weight or Body Fat Standards IAW AR 600-9.

Although each Cadet will have his or her own nemesis, the best way to receive the badge is to remain consistent, motivated, and calm, and to keep the desire throughout each event at LDAC. The easiest way to approach each event is one at a time and to practice with a buddy before testing. Do your best by remaining focused and put your best effort forth in all that you do and you will not be disappointed. In the end, you will have earned the prestige and met the challenge of Recondo but more importantly, you will receive a ½ point more on the National OML.

# 5 The Mission

A Cadet's mission at LDAC is simple...GET LDAC CREDIT AT ALL COST! This chapter is very short and is intended to be one page but don't underestimate the importance of this chapter because this is what a Cadet must do! If a Cadet understands the commander's intent then the Cadet can complete the mission. As stated before, GET LDAC CREDIT AT ALL COST is your PMS's intent. He is your commander so that should be easily understood. RIGHT! Of course there are some ethical issues that you must take into account and understand that AT ALL COST does not mean violating one of the seven Army Values or ignoring an ethical issue that may present itself. Finally, although an Excellent "E" Rating at LDAC will place a Cadet higher on the National Order of Merit List (OML), a Satisfactory "S" Rating or a Needs Improvement "N" Rating with LDAC credit gets the mission accomplished. Remember, there are certain tasks that must be accomplished before you gain LDAC Credit and they will be discussed within the next section of this book. Okay, let's write your mission statement for LDAC:

*Cadet (your name)________________________________ conducts individual and collective training during ROTC MS III Year in vicinity of (IVO) (your university)____________________*

*________________________________in order to (IOT) gain the necessary leadership characteristics to attend LDAC and receive LDAC Credit.*

## IN LAYMAN'S TERMS:

## **"GIT-R-DUN"**

# The LDAC Experience

## Lead, Follow or Get the HELL Out of the Way

# 6 Eliminating the Threat

Eliminating the threat at LDAC requires removing unknowns before attendance. One of the largest unknowns of LDAC is that Cadets don't understand what to expect or what is expected from them to succeed. Although this can be overwhelming and frightening to those Cadets who lack confidence, the best viewpoint is to realize LDAC is nothing more than a military training exercise planned by ROTC Cadre and executed by Cadets. Cadets are responsible for providing leadership and participating in the training. The next few pages will explain the intricate details of the LDAC concept, including the different phases and what you should expect.

As a prior Non-Commissioned Officer and Officer in Special Forces, the only three things I ever needed from my boss were Intent, End State and what Resources I had or could get to complete the mission. If you know the Intent and End State, you don't have to be told how to suck the egg, which allows you to accomplish the mission with the most latitude. However, Resources are another story and there are never enough—so get used to it! Most of the time, you have to do more with less and still accomplish your mission. Remember this basic concept when you become a Platoon Leader and don't be a Micro-Manager. The bottom line is to give your intent, desired end state and most important, let your subordinate leaders execute and you supervise. Now ask yourself three questions:

*1) "What is my Intent for attending LDAC?"*

*2) "What is my End State for attending LDAC?"*

*3) "What Resources do I have at my disposal?"*

By determining these three key pieces of information, you will be able to succeed. Think about what you want to achieve at LDAC; **THINK GOALS**! Most Cadets have the impression that LDAC is a useless and ambiguous training event that will not teach them anything. Some Cadets dread attending LDAC because they either fear the unknown, they listen to MS IV horror stories or they lack information from their Cadre about what to expect. If you allow yourself to succumb to these vulnerabilities, you may pass LDAC but you will not excel or be comfortable. Trust me; you have received the training necessary to succeed at LDAC. Now take the training to LDAC and use this awesome opportunity to apply your leadership skills in a military training environment without distractions. One of my main purposes for writing **"COLD HARD FACTS ABOUT LDAC"** is to educate Cadets in ROTC so that they reduce their fear of the unknowns at LDAC. Great leaders have confidence in their ability to make decisions, even in an unknown situation, therefore why shouldn't you. The bottom line is to apply the Golden Diamond Method at LDAC and make the best decision possible with the given information. There will be situations at LDAC that require a decision to be made in a timely manner; these decisions may be at critical times during your leadership and may be the difference between an "E" or an "S". Confidence comes from training and believe me, you will make mistakes during training; the difference between and "E" and an "S" is how a Cadet make the decision to revise the plan and how it is implemented. Therefore, embrace your training at ROTC and trust that you have been properly prepared to KICK ASS!

Most Cadets do not have a specific intent before attending LDAC but if you ask Cadets what their intent is, some will say to receive an "E" Rating, others will say to pass all events and receive LDAC credit and others don't have any idea. The Professor of Military Science (PMS) at your campus takes input from Cadet's Primary Instructor, Cadre, Field Training Exercises (FTX), MS IV Cadets and peer ranking to develop an OML ranking order and Cadet Evaluation Report (CER). Cadets receive the PMS's leadership assessment through an On Campus CER that is communicates an honest leadership assessment to include the Cadet's strengths and weaknesses prior to attending LDAC. The CER counseling session becomes an important milestone at the end of the spring semester because it gives you a good understanding about where you should focus your efforts before leaving for LDAC. The tools listed above allows the Cadre and you to develop and implement a supplemental training plan that addresses your deficiencies so you can develop a realistic Intent and End State that can be easily obtained at LDAC. To achieve this Intent and End State, you must set specific goals for each of the graded events that will determine your outcome. These goals will be discussed throughout different sections of the book.

**COLD HARD FACT 6:** Failure to make a decision is worse than making a wrong decision. Therefore, if a wrong decision is made, correct the mistake and Charlie Mike (Continue the Mission).

*Intent and End State example, "My intent is to lead when in charge and be a good follower when not in charge. Consistently improve each leadership position by listening to what is said during After Action Reviews (AAR) and Counseling. Complete all evaluated events with a first time GO and within the top 30% of my platoon. My End State is to receive LDAC Credit with an overall Satisfactory "S" Rating and 50% Excellent "E" LPIs."*

Key tasks are what you must accomplish to meet the Intent and End State. These tasks can be endless and varied but you must narrow them down to about five or seven that will encompass all the others or otherwise you will forget them if you make too long of a list. Examples of key tasks for LDAC preparation:

Participate in all individual and collective training events

Provide dynamic leadership through surety and confidence

Pass APFT

Pass Land Navigation

Self Assessment of Skills

## The LDAC Concept

The Leader Development and Assessment Course (LDAC) is the most significant training event that you will attend as an Army ROTC Cadet. The month-long commissioning course incorporates individual skills into a collective training environment designed to develop and evaluate ROTC Cadet's leadership potential. Although each regiment is standardized, the training can be very mentally and physically challenging depending on a Cadet's previous training. A Cadet's intelligence, emotions, stamina and common sense are tested each and every day by scenarios that are developed to confront a Cadet's ability to perform overwhelming tasks and make difficult decisions.

Prior to 2010, Cadet Command used an internal scenario that focused on a unit preparing to deploy from the United States to a country named Palomas that had been invaded by Cacueta. The new LDAC scenario is based off a TRADOC Common Training Scenario (CTC), Caucasus Region and was designed to meet leader development needs of institutional training and education within the Army. There are two approved scenarios that allow for planning and coordinating simultaneous and full spectrum operations in a joint decisive action setting within the Contemporary Operational Environment (COE). The CTS scenario used at LDAC encompasses full spectrum operations in the Caspian Sea Region. The scenario includes variables and units that enable LDAC to function as a COE without making fundamental shifts of training and guidance on college campuses.

The Caucasus Region Scenario has Cadets deploying to Intermediate Staging Base (ISB) Lewis to support combat operations against Arianan Insurgent Forces in Southern Atropia. Cadets have the follow on mission to conduct Stability Operations along the Atropia and Ariana international border to maintain Atropia's territorial integrity and prevent enemy forces from influencing the local population. This scenario's intent is to allow Cadets to experience procedures and necessary tasks a unit would conduct before deployment. The scenario allows Cadets to be placed into evaluated leadership positions for a twenty-four hour period. During this scenario, there are specific tasks that must be accomplished just as a military unit would accomplish prior to an overseas deployment. The entire scenario is Cadet-led and Cadre will not interfere unless life, limb, eyesight or a chance of transportation scheduling is threatened. If a Cadre interferes and takes charge of the situation, this is considered "Cadre Override" and you can bet your leadership evaluation will not be the best. One special aspect of LDAC is that each regiment is affiliated with a past unit.

As with any unit, situations occur each day that require leaders to assess, innovate, make sound judgment and get results by executing those decisions. What I have just described is the Golden Diamond and tells a lot about a leader. Of course, during each of the situations, the Cadet is constantly being observed for behavior. The Seventeen Leadership Performance Indicators (LPI) is used to measure the Cadet's behavior and give an honest assessment and evaluation. The scenario is executed in four-phases. Each of the phases has specific training that takes place and must be accomplished before moving to the next phase. Each phase builds on the previous phase and becomes more challenging. In the end, your platoon operates more and more as an effective team by traveling through each phase of the team building process. Phase I (Mobilization & Staging) consists of in-processing, physical examinations and individual training; this phase represents preparation for combat operations. Phase II (Onward Movement & Integration) Cadets move into a secure Assembly Area to conduct individual and collective training for small unit level combat operations. Training includes FLRC, Land Navigation, confidence course and first-aid. Phase III (Deployment into Theater) is the occupation of a Tactical Training Base (FOB) near the international border. During this phase, Cadets conduct Squad Situational Training Exercises (SSTX) and Platoon Situational Training Exercises (PSTX) in support of stability operations. Missions include reconnaissance, raids, ambushes and other combat operations. Phase IV (Demobilization) is recovery operations with an emotional roller coaster toward redeployment to Fort Lewis for retraining, final assessments and Graduation. During this phase you will conduct the famous Victory March back to the barracks. All the other regiments that arrived LDAC after you will be standing on the streets to cheer you on; they will be quietly jealous because you are just about finished. Finally, ensure you ask your Cadre to provide you the Road to War Briefing and the Caucasus Region WF GTA early in the spring semester so that you can be familiar with the significant events that led up to your deployment to Ft. Lewis, WA.

# 7 Individual Skills

## Army Physical Fitness Test (APFT) and Physical Fitness

Historically, Cadet APFT failure rate increases from those who attend the 7th Regiment and later. The most probable cause for this increased failure rate is the amount of time away from a mandatory or structured PT program at their battalion between the spring semester and the time they arrive at Warrior Forge. At LDAC, if you fail any one of the three graded events, you will receive an automatic Needs Improvement "N" Rating. Although these Cadets may still receive LDAC credit, they often spin out of control into a "Death Spiral" and continue to fail additional events simply due to the fact they get down on themselves and are not motivated from this point forward. Once this happens, these Cadets are labeled as "Marginal Cadets" and are flirting with the danger of being sent home without LDAC credit. Cadets who fail the first APFT will have to retest before moving into the next phase of LDAC. Prior to the APFT, Cadets go through a height/weight screening and receive physicals. Fail to meet the height/weight and body fat standards and you just failed to receive the Recondo Badge.

The Army Physical Fitness Test (APFT) is administered at LDAC and counts for approximately 9.56% of the total points that will be used to determine your National Order of Merit List. The event consists of push-ups, sit-ups and the 2-mile run. Don't underestimate the APFT event at LDAC. Usually the ROTC Physical Fitness Training conducted at least 3 days a week at your home campus is not sufficient to excel during this emotional event. You must conduct additional physical fitness outside of ROTC and meet this challenge with a positive and aggressive attitude. All MS III Cadets hear the rumors and horror stories from their MS IV Cadets on how hard 2LTs grade the APFT at LDAC. The truth is, the APFT grading standards do not change and the only variable is YOU! The best way to mitigate the "so called" tough grading standards is to prepare yourself at a higher standard so you welcome the challenge. If you are physically fit and prepared, you will exceed the standard with ease; if you are not, you will be challenged and found wanting. The only way to meet this high standard is to train effectively and understand how to train. Below is the secret stuff to assist you in obtaining the 9.56% of total points:

A technique that I have used to take my APFT for years is to set goals and identify the number of push-ups and sit-ups that I want to accomplish at different times throughout the two-minute testing of each event. Even though each Cadet has different physical attributes and abilities due to his or her previous training and preparation, the one common goal that each person must have, is to do the best in each event. In order to perform well on game day, proper preparation is required on a routine basis. A routine that has been proven is to perform push-ups and sit-ups before you go to bed at night and when you wake up in the morning. The number of push-ups and sit-ups will depend on your level of physical fitness and should increase to a level where you can perform 40% of your maximum push-ups and 50% of your maximum sit-ups every night. Practice the push-ups and sit-ups in front of a mirror or use a video camera to ensure the proper form is being conducted; most people try to perform the push-ups too quickly with the mindset of getting them done, instead of performing them correctly.

**Leadership Insight 6:** If your goal is to get 60 points in each event, **YOU WILL FAIL**! Always do your best in each event during every APFT. Don't let your enemy, Tiny Heart Syndrome, win the battle!

**COLD HARD FACT 7:** Failure to do PT after completing the spring semester will deteriorate your physical fitness and you will FAIL the APFT. Fail the APFT twice and you are going home without LDAC credit.

As with any graded event, there are crucial times during the push-up and sit-up events that can motivate you to obtain the desired goals. The first crucial time is at the 1-minute mark. I maximize my performance by getting into a zone and forcing myself to make a breath at each repetition. For push-ups, I exhale a short breath at the extension of my arms. For sit-ups, I exhale a short breath at the up stage of the sit-up. This allows me to not worry about the number and to get into a rhythm. Once you arrive at your goal, take a rest and then break down the remainder of the repetitions to a number that you can obtain before time expires. Even if you max your event, continue to use the entire 2-minute time frame to squeeze out those final reps in order to improve your stamina and train your muscles to perform when they are fatigued. When doing push-ups and your muscles become fatigued, move your hands to different positions so that you use different parts of the muscles. For example, face your fingers at a 45-degree angle outward or a 45-degree angle inward so that different triceps muscles are used. This should allow you to obtain a few more repetitions.

You must learn how to train correctly before you will be able to maximize your true potential. Running 2-miles daily does not afford you this opportunity. You must have a training program that incorporates, Long Slow Distance (LSD), tempo runs and interval training. Additionally, the 2-mile course must be broken down into known distances called legs so that you have a reference point to measure your performance. I recommend dividing the course into ½ mile segments. Most courses are marked every ½ mile, which will allow you to judge your performance. For example, if your goal is to run the 2-miles in thirteen minutes, then you want to break down the course into 4 segments. The first segment (at the ½ mile mark), would require the Cadet arrive at this marker no later than 3 minutes and 15 seconds, at the 1-mile mark, 6 minutes and 30 seconds, at the 1 ½ mile mark, 9 minutes and 45 seconds and finally arrive at the finish line at 13 minutes. With this method, individuals can gauge their performance during the run and adapt to meet their goal. Of course this is an example in a perfect world where the goal is to run each 1-mile segments with equal time. However, everyone is different and may not run the first mile as quick as the second mile. Each Cadet must determine his or her strengths and tailor them to a time standard. When I run the 2-mile course, I tend to start out slow and then adjust my speed to where I am finishing the last ½ mile at a much faster pace. This allows me to warm up and then finish strong. Below are the goals that I set to receive a score of 300 points on the APFT. At LDAC, more than likely, you will not have time to run the 2-mile course before taking the record APFT. But if at all possible, get out and go for a 20-minute feel-good run to familiarize yourself with the area and weather. Finally, STOP being a couch potato; defeat the Tiny Heart Syndrome by implementing a physical fitness plan. Below is a sample individual goal for each APFT event:

**Leadership Insight 7:** Don't allow the APFT at LDAC to become an emotional event and whatever you do, don't wait to start training for LDAC one week before you leave; my recommendation is to have a continuous physical fitness program throughout the year but if that is not possible, start at a minimum of four weeks before departing for LDAC. Your preparation begins during your first physical fitness workout at ROTC with a more focused approach and higher intensity level during the MS III year; form and correctness must not be compromised! Attend every PT session and add additional physical fitness sessions outside of ROTC. The best way to prepare mentally and physically before arriving at LDAC is to have a plan for how you will train and get the most out of your performance. Run with someone faster than you for motivation and moral support.

**COLD HARD FACT 8:** If you fail an APFT event, you are an automatic "N" at LDAC. The APFT is a MUST pass to get LDAC credit. Remember the National Order of Merit List – Higher Score = Branch Placement.

## Table 5: APFT Milestones

**Push-up (82 Repetitions)**

| | |
|---|---|
| 1 Minute | 60 Repetitions |
| 1 Minute 30 Seconds | 75 Repetitions |
| 2 Minutes | 82 Repetitions |

| Time | Repetition Goal |
|---|---|
| **Sit-up (100 Repetitions)** | |
| 1 Minute | 60 Repetitions |
| 1 Minute 30 Seconds | 85 Repetitions |
| 2 Minutes | 100 Repetitions |
| **2 Mile Run (12:30)** | |
| @ ½ Mile Mark | 3 Min 15 Sec |
| @ 1 Mile Mark | 6 Min 30 Sec |
| @ 1 ½ Mile Mark | 9 Min 30 Sec |
| @ 2 Mile Mark | 12 Min 30 Sec |

## Land Navigation

Land Navigation is a perishable skill that requires practice in order for Cadets to be confident and successful. This is the second most critical individual event that you will be tested on at LDAC. The event consists of 3 events: 1) 20 question written test 2) Land Navigation day testing and 3) Land Navigation night testing. All 3 events total a possible 100 points with a minimum of 70 points per event equaling a passing score.

During your MS III year, land navigation will be a major focus area for you and your ROTC training program. Some of you will get very proficient at this delicate skill after a couple of times and others will take multiple times to include a practice session the week before leaving for LDAC. Pay special attention to this section of the book because I will focus on the composition of each event and what you must do to be successful. Other valuable information will consist of route planning and tools to assist you in passing the course. I suggest that Cadets look at land navigation as if they were on a treasure hunt. The main difference is that you have the answer key at your fingertips to find the treasure (Markers) and no one is trying to kill you while you are looking for the treasure.

**Leadership Insight 8:** Fail one land navigation event and the best Overall Rating you can obtain is an "N." Although your PTO can migrate you to a higher mark, The Warrior Forge Commander views land navigation as an "Individual Capabilities Test" and doesn't see the need to migrate Cadets.

The **written test** is composed of 20 multiple-choice questions that require specific technical skills to get the right answer. You will have 90 minutes to answer 20 multiple-choice questions in a classroom environment. You will receive multiple versions of the test so don't be stupid and try to cheat because you will get caught and be released from LDAC with an honor and integrity violation and you will get the wrong answer anyway. Each question will be worth 1 point and you must answer 14 of the 20 questions correctly for a minimum score of 70%. If you are going for the "coveted" RECONDO, you must answer 16 of 20 questions correctly for a score of 80%. Finally, the reason Cadets do not pass the written land navigation test at LDAC is usually because they failed to do one of two things 1) They did not practice or review their procedures before leaving for LDAC 2) They did not conduct an error check before submitting in the test. Performing an error check is reviewing your calculations and test answers by using all available testing time. Error checking ensures the Cadet marked the proper response to the proper question and checks to see if the correct GM angle was used. Finally, your school should give you one last land navigation written test at least one week before leaving to ensure that you are tuned-up and ready to kick butt at LDAC-I-STAN.

Before attending LDAC, you will take a series of different land navigation tests at your campus that may or may not be multiple-choice. During my tenure as a MS III Instructor at the University of Central Florida (UCF) "Fighting Knights," I always gave practice tests that required the Cadet to write their answers on paper so that they could not guess the correct answer from a multiple-choice test. This is a great tool so that your instructor can identify your weak areas. Either way, more testing before attending LDAC will benefit you.

The written test can be comprised of a series of different questions ranging from terrain features, marginal data questions, determine road and straight-line distance, determining 6 & 8 digit grid coordinates, resection & intersection questions and azimuth calculations. The most difficult questions for Cadets seem to be: 1) determining road distance and 2) resection & intersection questions. These are a bit tricky but with the right procedures and practice you will be able to master these types of questions with ease. At LDAC, the desks are flat and give you sufficient room to take the test. A good website that sells software that will allow you to practice basic land navigation to include written questions is **www.550cord.com.** They have a great program that allows you to use the online demo version for free as many times as you want. Additionally, the Warrior Forge Western Region website has practice test for download.

**Leadership Insight 9:** When taking the Land Navigation Written Test, go through and answer the easiest questions first. Leave the more difficult questions (e.g. resection & intersection, road distance) to the end so if time becomes an issue, you can still receive a passing score. Ensure that you mark the correct question with the correct answer and never leave a question blank. If you are unsure of the answer, make an educated guess even if time is running out. Most of all remember, Hidden (Hill) – Valley – Ranch (Ridge) – Salad (Saddle) – Dressing (Depression) for the major terrain features.

**Land Navigation Day Testing** consists of a practical application exercise on a designated course that requires you to apply your understanding of the technical aspect of day land navigation. Cadets have 5 hours to complete the course and must find a minimum of 5 out of 8 points to pass with a score of 70%. If a RECONDO Badge is your goal, 6 out of 8 points are required with a passing score of 80%. Although the distances vary from point to point, Cadets will travel approximately 5 –8 kilometers during the Day Course. Failure to find 5 correct points in the designated time will result in not meeting the standard and require Cadets to retest after receiving corrective training.

**Land Navigation Night Testing** consists of a practical application exercise on a designated course that requires Cadets to demonstrate their understanding of the technical aspect of land navigation during hours of limited visibility. Cadets have 3 hours to complete the course and must find 3 out of 5 Land Navigation Markers without Chemical Lights to pass with a score of 70%. The RECONDO Badge requires finding 4 out of 5 points for a score of 80%. These distances will also vary from point to point with a total distance of approximately 4 –6 kilometers during the Night Course. Failure to find 3 correct points in the designated time will result in not meeting the standard and require the Cadet retest after receiving corrective training. For those Cadets afraid of the dark, GET OVER IT! There is nothing at LDAC that is going to attack you! The animals that are out in the woods are more afraid of you than you are of them. Concentrate on the pace count and move out and find your points because if you lallygag around or feel sorry for yourself, you will have to go back out in the dark and do the course again.

**Leadership Insight 10:** Established water points are the only points that have chemical lights attached to them. These sites are known points and can be used as an attack point to assist in finding your point. However, talking to other Cadets is not tolerated and Cadre is positioned at these locations to ensure cheating is not occurring. Cheating results in a "Go HOME" score.

Conducting land navigation at LDAC is not any different than any other land navigation course that you have done in the past. Cadets arrive at the land navigation site and receive final classes before negotiating the course. Prior to testing, Cadets verify pace count, verify their magnetic compass calibration and receive a land navigation orientation briefing from the Land Navigation Committee. During the briefing, Cadets sit in bleachers around a massive size terrain model that gives an overall view of the course. On the terrain model there will be several established water points (WP) and reference points (RP) that can be used to your advantage during the course, so pay attention and write down the grid coordinates. After the briefing, you will be able to plot these known points on your map, which you can use as attack points or reference points while on the course. The next FLAR is to use the known points! Once you begin the testing for either day or night, Cadets are called by name and given a course consisting of a start point and 8 additional points for day, 5 points for night. Once you are given your score sheet, the time will be recorded and you can plot your course and begin. Don't take more than 30 minutes to plot your course and move out. Time is crucial! Once you find your point, you will be required to write down a specific number or letter at the point and mark the scorecard with a clacker that has a specific pattern and design. Figure 7 shows the configuration of the clacker and the land navigation marker. It is very important that you do this because without both markings, you will not receive credit for that point. More importantly, don't forget that you have a specific amount of time to complete the course—5 hours during the day and 3 hours during the night.

**COLD HARD FACT 9:** Failure to mark the scorecard in the proper location with the clacker and document the point number or letter results in not receiving credit for that point. A lost scorecard results in a BIG FAT "NO GO" and a retest!

## Table 6 – Land Navigation Testing Summary

| LN EVENTS | TEST COMPOSITION | TESTING DURATION | MINIMUM REQ | MAX SCORE | MIN SCORE | RECONDO MINS |
|---|---|---|---|---|---|---|
| WRITTEN | 20 QUESTIONS | 90 MIN | ANSWER 14 QUESTIONS | 20 PTS | 14 PTS | 16 PTS |
| DAY TESTING | LAND NAV MARKERS | 5 HOURS | FIND 5 OUT OF 8 MARKERS | 50 PTS | 35 PTS | 40 PTS |
| NIGHT TESTING | LAND NAV MARKERS | 3 HOURS | FIND 3 OUT OF 5 MARKERS | 30 PTS | 21 PTS | 24 PTS |
| TOTAL | | | | 100 PTS | 70 PTS | 80 PTS |

### Necessary Equipment

In order to pass land navigation you must have the necessary tools to assist you and they must work correctly. Many Cadets do not conduct the Pre-Combat Checks (PCC) and Pre-Combat Inspections (PCI) on their land navigation equipment before leaving their home campus, which results in them being unprepared to negotiate the course and a possible fail. Although most schools require Cadets to do one final PCC / PCI with their instructors before leaving, it is ultimately the responsibility of the Cadet to notify the Supply Tech to replace malfunctioning equipment. Below are the bare essentials of land navigation equipment that you will need and what PCIs you must perform:

*Military Compass* – Ensure the bezel ring works, the face is not scratched and the numbers are visible. Verify that the north-seeking needle is accurate by verifying the reading from a designated known point and ensure that the illuminating dial works. Tie the compass off to your LBE so that you don't lose it. Don't forget that three clicks on the bezel ring equal 1 degree. While walking to find your point, ensure you alternate directions when bypassing an obstacle. If a Cadet consistently goes to a specific side of an obstacle to go around the obstacle, he or she will tend to drift and not come out directly on the point.

*Mechanical Pencil* – A .5mm mechanical pencil is a must! The smaller pencil lead allows you to make precise marks on the map that allows for accurate plotting and obtaining an exact azimuth. Ensure you take a back up or two so when you lose the first pencil it will not be an emotional event. A Sharpie is not a .5mm mechanical pencil!

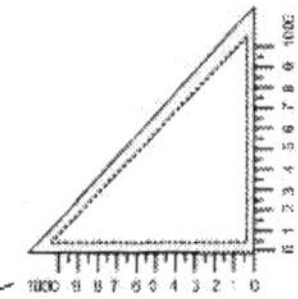

**Figure 8a:**
**Excess Plastic View**

**Figure 7:**
**Ranger Joes Protractor**

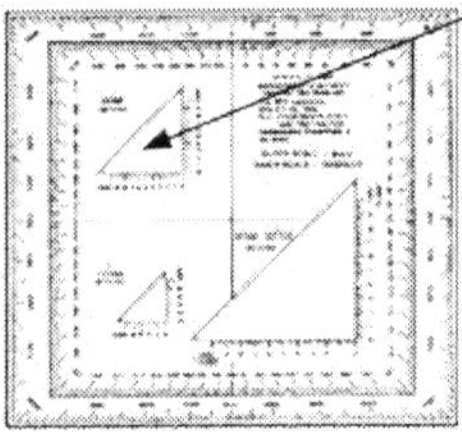

**Figure 8:**
**Army Protractor**

*Protractor* – Invest in two good protractors before going to LDAC. Ensure the protractors are scratch proof and have a production date later than Oct 2005. Ranger Joe's has the best protractor to take to LDAC, however, some TACs will not allow you to have it. This is one of those key pieces of equipment that is a must. For some reason, LDAC is living in the stone ages and forces Cadets to use the Army issued protractor. The Army isued protractor makes it 'very difficult to get an accurate grid because there is excess plastic inside the right angle of the scale. Figures 7 thru 9 show the difference between the two protractors to include the excess plastic that accompanies the Army Issued GTA 5-2-12. The Ranger Joe's protractor makes life easier and can be placed inside your hat for easy access. Of course, some Cadre will say this is an integrity issue so you have to make the choice. I chalk this up to the common sense rule and for some reason it has not reached the makers of the military issued protractor. If by chance they do not allow you to use the Ranger Joe's Protractor, modifications must be made to allow for accurate plotting. The first modification that can be done is to take a knife and shave the excess plastic off. Of course this takes great precision to not damage the protractor. The second modification is to offset the plotting of the points to make up for the excess plastic.

**Leadership Insight 11:** TACs very rarely check the inside of your hat, boots, or cargo pockets of the ACUs. (Hint, hint!).

*Flashlight* with fresh batteries – A flashlight with a red and white lens is a must! The point markers are red and white and are easy to find in the day but harder at night with a red lens flashlight. Many Cadets use headlamps, which make it more accessible. If you use a headlamp, do not walk with the headlamp on because it will affect your night vision, which makes it harder to see the markers. The headlamp is best used when you are in the vicinity of the point and allows for you to see a reflection of the marker quicker than with the naked eye. Currently, LDAC does not allow headlamps but maybe within the next couple light years, they will come up to speed. Then again some PTOs let their Cadets have them; good luck with that!

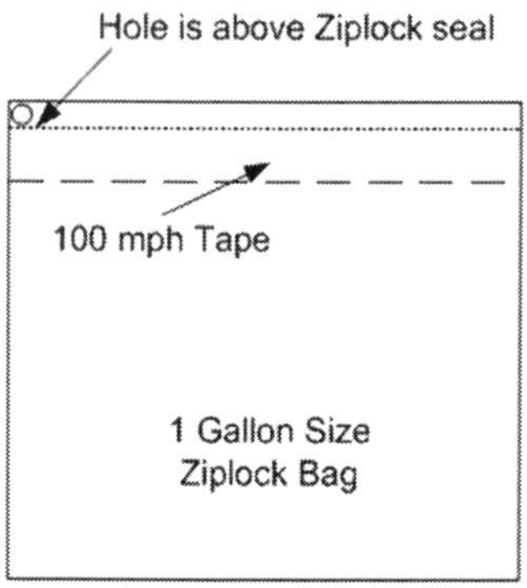


**Figure 9:**
**Expedient Map Bag Sketch**

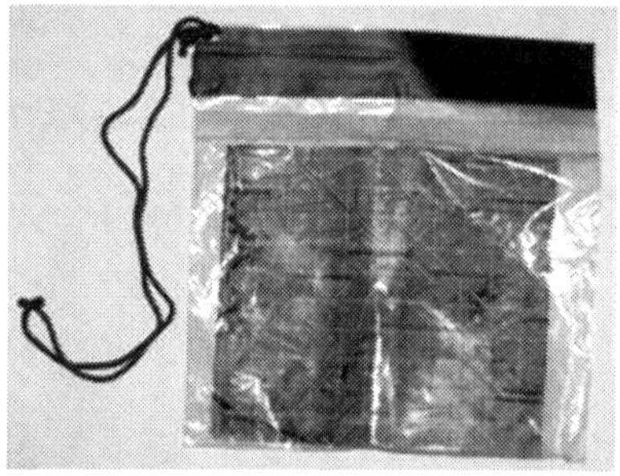

**Figure 10:**
**Expedient Map Bag Photo**

*Field Expedient Map Bag* – A pre-constructed map bag is not authorized at LDAC but from experience, it depends on your TAC if they will allow you to keep the map bag. If your TAC does not let you keep the map bag you bought from Ranger Joe's, you will still need one for those rainy days at Fort Lewis. Don't fret; you can make your own so that you can keep your map and score sheet dry. A field expedient map bag is just as good as one you buy but it costs a hell of a lot less. I used this same field expedient map bag when I went through Special Forces Selection and it never failed me! The field expedient map bag can be tied to your belt to ensure that your equipment is not lost on the land navigation course. The map bag allows you to have a waterproof location to place your scorecard, map, and protractor into one location so that it remains dry. This becomes another FLAR because if you lose your map or scorecard, you are pretty much dead in the water and will get a NOGO. Additionally, this is also a great opportunity to teach other Cadets at LDAC how to make a field expedient map bag (Positive Spot Report Opportunity). All you need is a 1-gallon size Ziploc Bag, 100 mph tape and 550 cord.

*Functional Watch* – A watch that works is a must not only for Land Navigation but for Warrior Forge! Make sure you carry a back up. Enough said!

**Leadership Insight 12:** Conduct an inspection of your equipment before leaving each point on the course after marking your scorecard with the clacker and appropriate letter or number.

## Time Management and Route Planning

Immediately after receiving the designated course with your points for day or night testing, you must manage your time wisely to finish the course on time. From the time you receive the points, you should make it a goal to plot your points and calculate necessary data (distance and azimuth) to find each point resulting in efficient route planning. This should not take more than 20 – 30 minutes. If you take longer than 30 minutes you are losing valuable time on the course that could be the difference between a GO and NOGO. The next step is to move to a location so that you can place your map flat on the ground. This will allow you to have a good foundation for plotting and the ability to lie over the top of the map. If it is raining, use your poncho! No matter what, even if it is raining, lie on the ground over the map so that you plot the points accurately. When plotting, make small marks and circle the point so that you can see it with ease. After plotting all the points, conduct a quick error check, determine a route that allows you to find all the points in a circular rotation so that you are not moving back and forth across the terrain. Next, determine the proper calculations for your route. I suggest going to the closest point first so you build your confidence and then move out from there. Ensure you put your data on a Route Planning Reference Sheet so that you can reference the data very quickly once you arrive at your next point.

*Route planning* is a very important part of being successful during day and night testing. A well-planned route will reduce your stress, the distance you travel and the amount of brush that you have to bust in order to get to your points. Many experienced land navigation personnel love to go cross country and do things "Old School" (dead reckoning) with the result of getting lost or exerting more energy than necessary. Don't be the Cadet who fails land navigation because you went "Old School" when it was not necessary. Although there are applications at LDAC for dead reckoning, Cadets must know when to use the right tool, at the right time and for the right situation. Oh yeah, don't forget, you are allowed to use the roads at LDAC! Although dead reckoning is a method for moving from point to point, I suggest using a combination. Use the roads as a major artery to cover the majority of the distance, establish attack points approximately 500 meters or less from the point you are trying to find, then use the dead reckoning method from the attack point to the point you are looking to find. The theory is that you use the roads to move quickly to an attack point and then use dead reckoning from the attack point to locate your point. The technique of establishing attack points for a reference point and then using dead reckoning to find the point is a great combination and allows you to find your points with precision, ease and minimum energy and time in the woods.

*Attack points* can be natural (intersection) or manmade (water points or reference points) and should be no more than 300 – 500 meters from the point you are trying to find. At LDAC, the water points and Reference Points are the most accurate points to use as an attack point because they were emplaced using a GPS. Attack point examples are intersections, fork in a river, hilltop, bridge etc. During plotting, designate an attack point close to the point you are going to find, then you can calculate the distance and direction from that attack point and then use dead reckoning (following your compass exactly for a specific distance) to go directly to the point. As stated before, an attack point should only be 300 – 500 meters away.

**Leadership Insight 13:** All the points at LDAC are approximately 400 meters from an attack point and were emplaced using a GPS for accuracy.

## Route Planning Information

The only authorized pre-formatted material or reference that is authorized at LCAC is the TACSOP, however, there is a loophole that allows Cadets to have additional notes by having hand written notes in their TASCOP prior to their arrival to include any addtional information written down in your note pad after arriving at LDAC. I suggest creating a Route Planning Reference Sheet so that you can effectively manage your route choice during land navigation.

The first step is to plot all of the points once you receive your score card. This should not take more than five minutes and must be done with precision. The second step is to determine the route sequence to your points which is a very crucial step! After determining the sequence, write down the order of how you will find the points and then calculate the azimuth and distance from start point to the attack point of the first point you are going to find. If you are taking roads (which you should be), you should write down the azimuth that the road follows to verify you are on the right road. Some roads on the land navigation course at LDAC are not accurately depicted or not on the map. Ensure you follow your map and are traveling on the right roads by identifying checkpoints along your route to track your progress to the attack point. I recommend going after the closest point first to gain confidence and to ensure the points were plotted correctly.

Next, calculate and write down the precise distance and direction from the attack point to the point. This is the information you will follow exactly by using dead reckoning. After taking the time to calculate this data, you must write it down on the Route Planning Reference Sheet before going on the course;

remember, NO MORE than 20 – 30 minutes to determine the necessary route planning information. The Route Planning Reference Sheet allows you to move from point to point quickly by referencing the sheet without removing your map and other items that could result in losing your score sheet.

Following is an example of how to prepare the Route Planning Reference Sheet to include the necessary data. Distances are from checkpoint to checkpoint in order to keep you on track while moving; maintain a pace count for accuracy. Most Cadets lose their pace count before reaching 500 meters so a set of Ranger Beads will assist in maintaining your pace count. Pay close attention to Step 2 because if you choose the correct route sequence, you have the latitude to return to the scorer's table and not have to find the last point if you are running short on time.

**Leadership Insight 14:** Take approximately 20 – 30 minutes to plot your entire route before moving onto the course. Ensure you transfer the data to the Route Planning Reference Sheet and then go find those points.

**COLD HARD FACT 10:** Cadets fail land navigation at LDAC for one of several reasons: 1) They do not keep an accurate pace count resulting in getting lost. 2) They are over confident and don't use their compasses. 3) They drift to the right or left because they stare at their compasses or continually go around obstacles in the same direction. 4) They leave their light on while moving from point to point which degrades their night vision capability. 5) They are terrified of the dark!

Following is a basic Route Planning Example with Reference Sheet to illustrate how information is gained and transcribed onto the Route Planning Reference Sheet. The Reference Sheet saves time and helps prevent losing scorecards, maps and other necessary equipment.

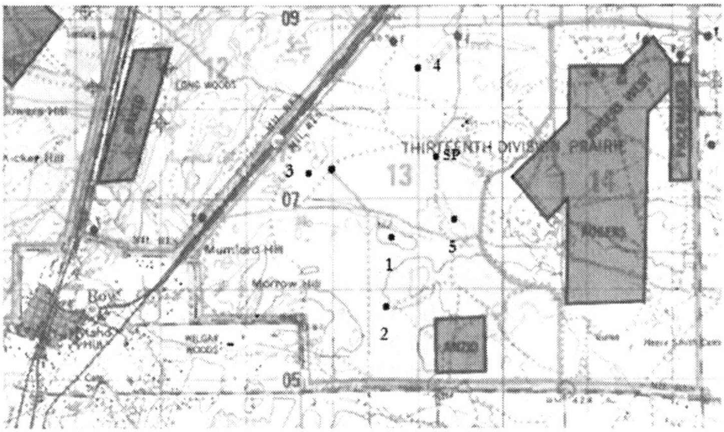


**Figure 11: Plot Points**

## Step 1 – Plot Land Navigation Points

## Step 2 – Determine Route
## SP □> 4 □> 3 > 1 > 5 > 2 >□ FP

**(other routes are possible)**

With the route sequence, shown in step 3, a Cadet has the opportunity to find point 5 and return to the start point if they feel they have found a sufficient amount of points to receive a

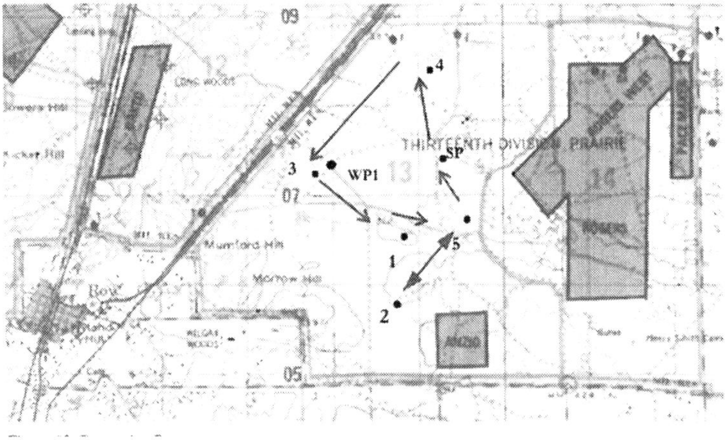


**Figure 12: Determine Route**

GO. This is particularly important if time becomes a factor. If the Cadet has enough time, they can easily go and find point 2 and then return to point 5 before returning to the start point. The bottom line is that you have options!

## Step 3 – Determine Checkpoints and Attack Points

In Step 3, the purple points signify the location of the Attack Point, the red arrow signifies the azimuth direction; the letter C signifies a checkpoint.

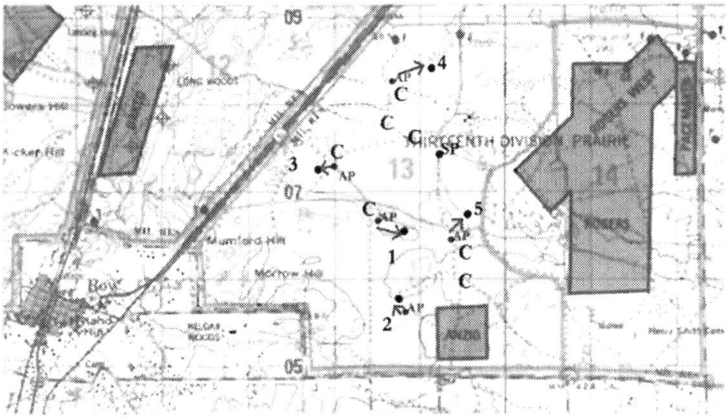


**Figure 13: Determine Attack and Check Points**

## Step 4 – Calculate Dist and Direction

### (always change grid azimuth to magnetic azimuth) before placing information into the Route Planning Reference Sheet (next page)

Finally, many Cadets will not walk directly up on their point because of drift or because they failed to follow their compass from the attack point to the marker. No matter what the case is, at this point it is necessary to find the point and not take a lot of time locating it. I recommend a couple of techniques to find the point. The first is to turn around and make sure you did not pass it; then look around the area 360 degrees slowly. More than likely it is very close and may be hidden by some vegetation. Next is to listen! There are other Cadets looking for the same point. However, I suggest not talking to other Cadets because there is Cadre at the markers, water and reference points who will turn Cadets in for not following instructions which could lead to an "Honor or Integrity Violation." If you can't find the point by stopping, looking and listening, continue along your azimuth for another 25 meters and mark your position by identifying an identifiable feature or using your PT belt. At that time, stop, look and listen again. If you still don't see anything, turn 90 degrees and walk 25 meters and repeat the stop, look and listen. If the point is not identified, return to your marker and do the same procedure in the other three directions from the base. If this does not work, continue another 25 meters and repeat the process. Finally, I would not spend more than 15 to 20 minutes looking for a point before I used another attack point or moved on to the next point. It is very easy to waste time at one point and then not make it back before time expires.

## Table 7 – Completed Route Planning Reference Sheet

| POINT | GRID | TO | GRID | ROAD CHECKPOINTS | ATK PT DESCRIP | AZIMUTH TO PT | DISTANCE TO PT |
|---|---|---|---|---|---|---|---|
| SP | ET 3922 0767 | 4 | ET 3894 0850 | Y-Int @ 375 m; Y-Int @ 125 m; T-int @ 250 m; 4-way Int @ 350 m; curve @ 200 m; curve @ 300 m | CURVE | 93° MAG | 250 M |
| 4 | ET 3894 0850 | 3 | ET 3776 0735 | Reverse checkpoints from above but turn R @ 2nd Y-Int @ 1,225m; WPI @ 4-way Int | WP 1 | 251° MAG | 240 M |
| 3 | ET 3776 0735 | 1 | ET 3861 0582 | Rtn to WPI and travel SE for 600 m | 4 WAY INT | 123° MAG | 325 M |
| 1 | ET 3861 0582 | 5 | ET 3942 0672 | AP and travel E/SE; Y-Int @ 600 m | Y-INT | 17° MAG | 175 M |
| 5 | ET 3942 0672 | 2 | ET 3861 0582 | AP and travel E/T-Int @ 150 m; T-Int @600 m; curve @ 700 m | CURVE | 250° MAG | 125 M |
| 2 | ET 3861 0582 | SP | ET 3922 0767 | Rtn to same AP that you used for pt 5; curve @ 375 m | | | |

## PLEASE COME FIND ME

Figure 14: Point Marker with Checker

**Land Navigation TA13 Score Card**

NAME:____________________________________

REGIMENT:______________COMPANY________

PLATOON:__________________________________

START TIME:______________________________

FINISH TIME:_____________________________

COURSE (DAY/NIGHT)_______________________

LANE #______________DATE:_________________

TOTAL VALUE OF POINTS:_____________________

PENALTY POINTS:___________________________

FINAL SCORE:_____________________________

| Letter/Number | Letter/Number |
|---|---|
| A / 10 | |
| Stamp | Stamp |
| Letter/Number | Letter/Number |
| Stamp | Stamp |
| Letter/Number | Letter/Number |
| Stamp | Stamp |
| Letter/Number | Letter/Number |
| Stamp | Stamp |

Figure 15: Sample Score Card

# 7 Garrison Operations

The definition of Garrison is a fortified military post that is permanently established and has troops stationed. In the military, soldiers are assigned to units, units are assigned to commands and commands are assigned to posts. All units in the military conduct Garrison Operations and for the most part, can be administrative or training related and are conducted a majority of the time. Garrison Operations at LDAC place Cadets at a military post to measure a Cadet's ability to lead in a non-threatening environment. Therefore there is very little threat of an actual enemy soldier from Caqueta (OPFOR) attacking you while in Garrison Operations. However, this is the most likely time you will be overwhelmed by another enemy force such as THS, Elusive Evaluators or The Inner Villain, which will result in the Cadet's ability to excel.

Garrison Operations consist of many different missions and tasks that must be accomplished in order for your unit to complete the required training before deploying into Palomas. The missions range from Physicals, APFT, Land Navigation, Basic Rifle Marksmanship, CWST, FLRC, US Weapons, Individual Tactical Training (ITT), Foot Marches, Obstacle Courses, drawing and turning in weapons, Hand Grenades, Rappelling, Branch Orientation, and Nuclear, Biological and Chemical (NBC) Training. You will have to identify specific and implied tasks during Troop Leading Procedures that must be completed in order to accomplish these missions. All of these training events or operations will not be accomplished in one day by one person but instead, everyone in your platoon will accomplish the tasks by rotating leadership positions every twenty-four hours. The key to excelling at Garrison Operations is to be organized, flexible, able to delegate and multi-task to the lowest level of leadership. I cannot emphasis enough the requirement to delegate, supervise and assess tasks or you will fail! Remember the Golden Diamond process. Enough said; use your subordinate leaders. The remainder of this section will give the leadership insights to techniques, missions and tasks that Cadets experience during Garrison Missions.

**SEA-TAC and In-processing**

The Army provides the method of transportation to and from LDAC depending on your location in reference to Fort Lewis, WA. Very few Cadets will be able to drive to LDAC unless they are within a reasonable driving distance to Fort Lewis, or it is cost effective. Once you arrive at Seattle-Tacoma (SEA-TAC) Airport, there are signs that direct you to a Warrior Forge Reception Area. Once at the SEA-TAC LDAC reception area, your name will be checked off a list and you will be placed on a bus to be transported to Fort Lewis. Upon arrival at Warrior

Forge, Regimental Cadre will direct you to an in-processing location where you must present your Military Identification Card, ID Tags, Orders, Airline Tickets, Medical Records, APFT Scorecard and any other follow-on information for CTLT or Cool Guy Schools such as Airborne or Air Assault. This is an important event because Cadre form perceptions of Cadets as soon as they arrive. These items are required and if you don't have them, you have just started out on the wrong foot. The way a Cadet dresses at this initial event is crucial. Those Cadets who arrive without a military haircut, wearing flip flops, surfer attire or dress like they are going to the club will be labeled. Females, those skimpy little skirts and shorts will get you noticed by the male species at the airport, on the plane and while picking up your luggage but after that, you get identified and placed into a category that is negative. Therefore, show up in respective attire that displays conduct becoming of a future officer in the U.S. Army. After in processing, you are assigned to a platoon and then you go through a shakedown to ensure that Cadets have the required equipment and do not have any prohibited items. When I was a PTO, my PTNCO and I conducted the inspection for the first group of Cadets and then they were put in charge for the remainder of the inspections under our supervision. As I stated before, my philosophy is that you are in charge and I am there to evaluate your leadership. Those individuals that are placed in charge during the first couple of days are evaluated on their leadership capabilities, which result in them receiving a "Spot Report." However, these Blue Card Evaluations have counted for Garrison Leadership Positions if there are not enough rotations to go around. Bottom Line, do your BEST every time because you never know when it is a Spot Report or an Evaluation.

During the inspection, all prohibited items are confiscated, missing items are documented, high dollar items are inventoried and secured and vehicles are inspected, locked and placed in a specific holding area while attending LDAC. Prohibited items at LDAC are map cases, Gortex, weapons or ammunition, knives with blades longer than 3 inches, cell phones and cash over $100 USD. Realistically, it is possible to sneak prohibited items into LDAC. However, when I was a PTO, my standing policy was Cadets are big boys and big girls and if they get caught with a prohibited item, they will have to deal with the repercussions. A key note is that every regiment usually relieves a Cadet from LDAC because they kept their cell phone. Don't let it happen to you; do the right thing. Finally, if a Cadet has a prescription for medication, the medication must be in its original container and have the Cadet's name on the container. If the medication is not in the original container, the medication will be confiscated.

**Leadership Insight 15:** Some Cadre is not aware of all the prohibited items, which result in some items not being taken during inspection.

**COLD HARD FACT 11:** Get caught with prohibited items after the in-processing inspection and you may be booted from LDAC for an honor violation. Fail the urinalysis at LDAC and you will be booted from Army ROTC.

## Warrior Forge (LDAC) Rules and Policies

There are certain rules and policies established at LDAC to ensure Cadets and Cadre remains on an even playing field. Although there is an in-depth written policy laying out these rules, this book goes in-depth on the crucial rules and policies that directly affect your leadership abilities. The intent is to give Cadets heads-up information to avoid any surprises about Cadet treatment, the use of alcohol and tobacco, visitation rights, profanity use, vehicle use, Post Exchange (PX) and phone use.

Once at LDAC, Cadets will be treated as adults and as future Military Officers. Cadre will assume that you have been properly trained prior to your arrival to Warrior Forge and understand the customs and courtesies of the military; therefore, your actions should resemble those of an adult and a Cadet pursuing a professional career with the military. In the military, Officers receive a certain level of respect due to the rank structure whether they deserve it or not; however, at LDAC, reality sets in and true respect must be earned despite an individual's rank! The respect from your peers must be earned every time you are in a leadership position to be an effective leader and make your time enjoyable at Warrior Forge. This will be determined by the way you carry yourself and how you convey information. If you are one of those high-speed potty mouths, you are in for a rude awakening because the use of profanity is not tolerated! Profanity is considered degrading and it offends others so refer back to Consideration of Others

Training prior to attending LDAC and remember that it is not all about you. Thus there are no double standards and Cadre is also prohibited from using this low form of communication. Additionally, Cadre are not allowed to haze, yell, or touch Cadets in a degrading manner, humiliate Cadets in front of their peers or have them do pushups for punishment.

Another restriction is the use of any tobacco products and alcoholic beverages at Warrior Forge. Cadets are not allowed to have tobacco products at any time and Cadre are not allowed to have tobacco products in front of Cadets. The consumption of alcoholic beverages is strictly forbidden, including any bars, clubs and facilities where the primary business is the sale of alcohol. Just for the record, Warrior Forge is defined as when a Cadet departs the point of origin (e.g. his/her home of record or university) until the Cadet has returned to his or her final destination after completing LDAC (e.g. home or record or point of destination) so no drinking in the airport bar on the way out to Fort Lewis. Don't worry, there will be plenty of alcohol and cancer producing devices after you receive credit for LDAC and return to your home campus. One final restriction that can end your ROTC and Army career is when Cadets are pressured or voluntarily take a few hits of a little illegal substance called marijuana. This is DOWN RIGHT DUMB because there is a urinalysis conducted at LDAC which results in one or two Cadets sent home every year because they come up hot. Afterwards, they are dismissed from the program. This is

DOWN RIGHT DUMB!

Visitations while at LDAC are not permitted since they cause distractions unless prior approval has been granted by the Commandant of Cadets. This includes Cadre who are at LDAC from your university in different regiments. If they do visit a Cadet, then the Cadet is more than likely having issues and needs to see a friendly face for some motivation. If you are receiving your commission at the end of Warrior Forge, you will not be able to visit your family members until the night prior to graduation, but you will be able to talk to them before hand because phones are returned prior to graduation to make phone contact and final coordination with them. Privately Owned Vehicles are off limits after you in-process until you graduate. This includes riding as a passenger or operating them. Finally, you are restricted to the regiment area unless the Cadre allow you to the leave the regiment area. If you are granted permission to leave, don't forget to have a buddy and sign out at the regimental headquarters, even during those very rare times that physical fitness can be conducted. Ensure you let your chain of command know where you are, including the run route.

There are two major milestones during Garrison Operations when referring to rules and policies. The first is D3. Prior to D3, you will conduct in processing, have the use of your phone and be required to call home or the university to inform them that you arrived safely. Cell phones must be turned in to the Cadre prior to D3. When D3 arrives, Cadets will experience one of the most emotional events at LDAC, the APFT.

The second major milestone is around D8. Prior to D8, Cadets are not allowed to talk in the Dining Facility and are not allowed to use the PX or the pay phones. If there is a requirement to go to the PX prior to D8, Cadets must be escorted by a Cadre member. After D8, these rules are removed from the Cadets. For Cadets, this is a bittersweet moment. Bitter because leadership just became more complex since Cadets who are not in leadership positions will want to go to the PX and phones instead of completing necessary tasks. These two privileges become training distracters because Cadets will not manage their time; they will become a Blue Falcon (Bravo Foxtrot) and fail to accomplish the necessary tasks. (Blue Falcon - Military synonym or code word for "Buddy F**ker". The guy who gets everyone else into trouble, sells out all his friends to higher ups, gets out of his duty, or gets away scotch free as all his friends get blamed. Military police are usually referred as Blue Falcons by other military occupational specialties. A socially acceptable reference to the acronym 'BF', also known as 'buddy f**k'; referring to an action that leads to 'f**king over' your 'buddy'. Common in military parlance.)

Once again, this will become very apparent to Cadre during Peer Evaluations. The sweet part of the moment is because Cadets are allowed to call family and friends and get "Pogey Bait" (chips and snacks) from the PX. Remember "Pogey Bait" must be consumed because it is not allowed in the barracks. One final note is that since 9/11, LDAC has had to make significant changes each year due to resources and budgeting so the D3 and D8 milestones may change from year to year.

**Leadership Insight 16:** Example of a Daily Activities Garrison

## Table 8: Daily Garrison Activities

| ACTIVITY | TIME | REMARKS |
|---|---|---|
| WAKE UP | 0500 | VARIES DEPENDING ON TRAINING |
| FIRST CALL | 0530 | VARIES DEPENDING ON TRAINING |
| PHYSICAL TRAINING | 0535 | CONDUCTED LESS THAN 5 TIMES AT LDAC |
| BREAKFAST | 0630 | CADET LEADERSHIP DETERMINES PLT ORDER |
| FORMATION | 0745 | CADET LEADERSHIP DETERMINES EXACT TIME |
| TRANSPORTATION TO TRAINING | 0800 | ADJUSTED BY TIME/DISTANCE TO TRAINING SITE |
| TRAINING | 0830 | |
| LUNCH | 1200 | VARIES |
| TRAINING | 1230 | |
| DINNER | 1715 | VARIES |
| REGIMENTAL OPORD PROCESS | 1800 | OR 30 MINUTES AFTER DINNER MEAL (LEADERSHIP) |
| PREPARATION FOR NEXT DAY | 1800 | |
| LIGHTS OUT | 2200 | THIS TIME **NEVER** CHANGES |

**COLD HARD FACT 12:** If a ROTC Cadre from your home campus makes a visit to you at LDAC, it is because your performance is lacking and you are considered an "At Risk" Cadet.

### Garrison Leadership Positions and Rotations

While conducting Garrison Operations, there are multiple leadership positions that Cadets fill to receive evaluations. Garrison leadership positions can be platoon leadership that are rotated daily amongst the platoon and company leadership that rotates between platoons assigned to the regiment. For example, if a company with the regiment has four platoons, your platoon will provide company leadership every fourth day. The platoon leadership comes from your platoon, which is divided into four squads with up to twelve Cadets per squad. The Company Commander (CO), Executive Officer (XO), First Sergeant (1SG), Platoon Leader (PL), Platoon Sergeant (PSG) and Guidon Bearer are filled with Cadets from each of the four squads. Company and platoon leadership can be drawn from any of the squads but the Squad leader (SL) position is filled internally from each assigned squad. In my experience, Cadets who are properly prepared will have a better company or platoon level leadership rating than a squad level leadership rating. A Squad Leader rating is the lowest level of leadership at LDAC and for the most part has the minimum number of moving parts to manage. In order to get an "E" Rating for a Squad Leader position, you have to be on your game by being proactive and thinking ahead to support your platoon leadership. If you are not forward thinking and busting your butt to support your leadership, forget about the "E" rating. Usually each Cadet will receive one leadership evaluation at the squad level and one evaluation at the platoon level or higher during Garrison Operations.

The leadership rotation at LDAC during Garrison Operations is twenty-four hours and usually spans from approximately 1800 to 1800 hours. Cadets are given a minimum of three to four days' notice before assuming leadership, including what leadership position is to be held so that Cadets have the maximum amount of time to prepare mentally to take charge. The leadership rotation is posted in the barracks **BUT** it is the Cadets' responsibility to monitor the list for changes. **THERE WILL BE CHANGES!** For example, Cadets will fail graded exercises and be required to retest especially for the most devastating event, Land Navigation. During this period of time, the second and third orders of effects cause a change in the leadership rotation and become a Cadre and Cadet's nightmare. The second order effect creates an issue for the PTO because they have to change the leadership rotation numerous times if a Cadet has to retest an event. The reason for the change is to give Cadets more rest so they have the most optimum situation to lead. If Cadets have to retest, they do not have the same awareness about leading for the full twenty-four hour time period since they are removed from the platoon and taken back to the training area. This is not fair to the Cadets or their peer leadership so the Cadet is removed from the rotation and placed in leadership at a later date. Therefore the best solution is for a Cadet who retests an event not to be scheduled for a leadership position the day following a retest to allow for maximum recovery. The third order of effect creates an issue for the Cadets. Once the leadership date changes, they have to start preparing for a different day's events. This screws over the proactive Cadets because they have been looking at the training schedule for the upcoming leadership rotation and started to write their OPORD; by the way, this is the preferred technique and pays big dividends if you are proactive. This is easily attainable because the training schedule is posted and the major muscle movements can be planned before receiving the company level OPORD. If the leadership rotation changes, be prepared to work with a different peer leadership group. It should not be a big deal to adjust the events and continue to prepare your OPORD. There are many different variations to inform Cadets of upcoming leadership positions but it is up to the PTO on what method to use. On the next page is an example of how leadership rotations may be identified.

## Table 9: Leadership Rotation Schedule Example

| **LEADERSHIP POSITION** | **MON. – D7** 25 JUN | **TUES. – D8** 26 JUN | **WED. – D9** 27 JUN | **THURS. – D10** 28 JUN |
|---|---|---|---|---|
| CO | CDT STYKER | | | |
| XO | CDT MRAP | | | |
| ISG | CDT HUSKEE | | | |
| GUIDON BEARER | CDT HUSKEE | | | |
| PL | CDT WARNO | CDT OPORD | CDT FRAGO | CDT METT-TC |
| PSG CDT SUPPORT | CDT SITUATION | CDT MISSION | CDT EXECUTN. | CDT SERVICE |
| I SL | CDT SUPPORT | CDT COMMAND | CDT SIGNAL | CDT COMPASS |
| 2 SL | CDT MAP | CDT PROTRCTR. | CDT PENCIL | CDT SALUTE |
| 3 SI | CDT ACTION | CDT LOCATION | CDT UNIT | CDT TIME |
| 4 SL | CDT EQPMNT. | CDT AMMO | CDT RIFLE | CDT PISTOL |

Following are the evaluated leadership positions with duties and responsibilities. Pay special attention to these duties and responsibilities because your Evaluator will use them as a guide to determine your overall assessment.

**Company Commander**

- Coordinate with Cadre Company Evaluation Officer (CEO) and Company 1SG
- Lead Cadet Company Leadership Team (CO, XO and 1SG)

- Conduct Troop Leading Procedures (TLP) at the company level; develop WARNO, OPORD and FRAGO
- Develop Risk Assessment, Transportation and Load Plan for company
- Organize, control and move the company to required locations
- Properly and efficiently utilize and supervise subordinate leaders
- Ensure safety is incorporated into training

**Executive Officer**

- Coordinate with Cadre CTO / 1SG for supplies and equipment to support training
- Perform assistant CO duties as required (e.g. brief movement plan or site occupation plan)
- Act as the CO in absence of the commander
- Coordinate and follow up with adjacent unit operations or with Cadre
- Conduct recon for company level operations (e.g. foot marches, APFT location)
- Ensure Safety Plan is executed
- Establish Load Plans for movement

**First Sergeant**

- Organize and hold company formations
- Ensure Platoon Sergeants (PSG) submit Personnel Status (PERSTAT) Reports
- Prepare and submit morning PERSTAT Reports; company accountability
- Control Sick Call and account for all company absences
- Control movement of the company
- Supervise training and field mess operations
- Perform other duties as the commander may direct
- Ensure safety is addressed and enforced
- Supervise Platoon Sergeants (PSG) in executing the commander's plan
- Ensure PSGs receive updates to OPORD

**Platoon Leader**

- Coordinate with the CO and XO
- Receive Company WARNO, OPORD and FRAGO
- Complete all company specified task from company OPORD
- Conduct platoon recon

- Conduct TLP at platoon level; develop and issue WARNO, OPORD, FRAGO
- Supervise and spot check PCC/PCI
- Conduct Rehearsals
- Develop platoon Risk Assessment, Transportation and Load Plan
- Think ahead; be prepared for the next phase of your OPORD

**Platoon Sergeant Responsibilities**

- Assist Platoon Leader with TLPs
- Coordinate with company 1SG and Squad Leaders
- Supervise squad leaders in execution of the platoon OPORD to include assigned task
- Be responsible for platoon PERSTAT to 1SG and accountability of platoon personnel
- Conduct PCC/PCI at platoon level
- Hold platoon formations and move platoon as necessary
- Supervise platoon opportunity training plan
- Ensure safety is addressed and enforced

### Squad Leaders

- Support PSG and PL
- Be responsible for squad PERSTAT to PSG and accountability of squad personnel
- Conduct TLP at the squad level; develop and issue WARNO, OPORD and FRAGO
- Ensure squad accomplishes all assigned task to include receiving equipment and supplies
- Supervise Team Leaders in execution of tasks
- Conduct PCC/PCI at squad level
- Hold squad formations and move squad as necessary
- Conduct squad opportunity training in accordance with platoon training focus
- Ensure safety is enforced

## Regiment Task Organization

During your spring semester, each Cadet will be assigned to a specific regiment at LDAC that will have a task organization comprised of two companies with three to four platoons in each company and four squads per platoon. There are numerous Cadres that will support the regiment and will arrive approximately seven days before the Cadets. There are approximately forty Cadres per regiment but for the most part Cadets interface with the Regimental Chain of Command and the Cadet Evaluators. Each regiment is commanded by a LTC as the Regimental TAC Officer (RTO), Regimental Executive Officer (RXO) and a Master Sergeant (MSG) or Sergeant Major (SGM) as the Regimental SGM. Additional Cadre at the regiment level will be the Operations Lieutenant (S-3) and NCO. Each company will have a Company TAC Officer (CTO), a Company Evaluation Officer (CEO) and a Company Evaluation NCO (CENCO). Each platoon will have a Platoon TAC Officer (PTO), Platoon TAC Lieutenant (PTLT) and a Platoon TAC Non Commissioned Officer (PTNCO). The CEO and CENCO write evaluations for company leadership positions and the PTO, PTLT and PTNCO write platoon leadership evaluations. The S-3 will issue out the regimental daily OPORD.

During Garrison Operations, there will be many tasks that must be accomplished to have mission success and receive a good evaluation. Cadet leadership are given specified task in the OPORD such as conduct Pre Marksmanship Instruction (PMI), provide Cadets to serve breakfast, draw weapons or conduct physical fitness training via the Regimental OPORD. Implied Tasks are tasks that you must accomplish in order to complete a specified task or your mission. Although many Cadets overlook the implied tasks while planning, these tasks demonstrate that the Cadet Leadership can understand concepts by conceptualizing what must be accomplished. As stated in Chapter 3, this is the difference between an "E" and an "S." Failure to identify a necessary implied task could make or break your leadership evaluation. Examples of implied tasks are to develop an operations order, coordinate for weapons and training aids for Pre Marksmanship Instruction (PMI), ensure all Cadets have weapon cards, develop a physical fitness-training plan, wall locker security, and barracks maintenance. Finally, specified tasks drive implied tasks and both of these must be tracked by the platoon sergeant.

**Ankle Biters**

My definition of an Ankle Biter is the minuscule details of an operation that must be addressed in grave detail . Failure to address these details can create a pesky little creature that annoys you until the end of an operation. Although these pesky little creatures allow the operation to continue, they slow down the efficient implementation considerably. If a Cadet allows three or more ankle biters to surface in a single military operation it can transform into a 10,000 pound gorilla that sits on the Cadet's back and pounds on him or her with the possibility of bringing the mission to a standstill. Final result is that the Cadet's leadership evaluation goes from an "E" to an "S" or an "S" to an "N"! Ankle biters are skilled professionals and are masters of camouflage but over the past twenty years, Warrior Forge has revealed several of the Ankle biter disguises. The most common are drawing and turning in weapons from the arms room, barracks maintenance, conducting training and developing a simple and feasible physical fitness and transportation plan. Drawing and turning in weapons is a tedious process that is accomplished on a regular basis at LDAC. Each Cadet must have a weapons card that contains the Cadet's information, including the weapon serial number. Each Cadet will draw a specific weapon with a serial number on it and that Cadet must turn that weapon back into the arms room in order to get the weapon card returned. The arms room has a procedure for tracking each weapon so if a weapon becomes missing it can be narrowed down to a specific Cadet and serial number. Cadets will be in line waiting for their turn to draw or turn-in a weapon which presents an opportunity to get in trouble for talking or playing "grab ass" which usually results in a negative spot report. A motivated platoon that properly planned for weapon's draw and turn-in can execute this task in approximately twenty minutes. A platoon that does not have a plan can take well over an hour. I highly suggest company leadership do a recon and ask the arms room Cadre for recommendations. Finally remember step 8 (Supervise) of the Troop Leading Procedures!

Barracks maintenance and the security of wall lockers always give Cadets a problem. Cadet leadership rarely assigns a squad the task to perform barracks maintenance and those who do, NEVER spot-check the barracks. Spot checking the squad in charge of cleaning the barracks allows leaders to verify the task was completed and ensure your fellow Cadets locked their wall lockers and secured their personal items. Failure to lock your wall locker results in a negative spot report the majority of the time.

If you are in leadership when physical fitness is scheduled, ensure you have a plan and the plan is executed. This is a great opportunity to get a lot of positive marks on the leadership dimensions. If the training schedule has one hour dedicated for physical fitness, then use the full hour. Failure to develop a plan or use the full hour is not a good thing for the leadership. Physical Fitness training is a great time to receive a positive spot report for calling cadence and showing leadership to your peers if you took the time to learn some cadence. My suggestion is to learn three running and three marching cadences before attending LDAC. Knowing these cadences and being able to relieve your buddies will show you are a team player, motivate the platoon and put you in a position to receive a positive spot report.

Opportunity training during the training day is a must and will assist you in a great squad leader evaluation. Each day, the training schedule shows the training focus that is nested toward the next few days training events. Therefore the implied task is to give out the training focus when developing the OPORD. If you are the platoon sergeant, supervise the squad leaders and ensure they are conducting the training. Finally if you are the squad leader, conduct the training at all costs. There will be time during the day (e.g. PTO time) to make the training happen and support your platoon leadership.

Transportation plans are non-existent during the first week of LDAC, because Cadet leadership fails to think about asking what types of vehicles are used to transport them to the training areas. Well, the answer to the question is not classified so to give you a heads up, LDAC use buses and a boxed in trailer on wheels called a Silver Side to transport Cadets to training areas while at LDAC. For the most part, there are three different configurations: 1) One silver side and a bus 2) Two buses or 3) Two Silver Sides. Although the configuration does not matter when developing a plan, the most important point is that your squads know which vehicle to load. The best option is to conduct a recon or ask the Cadre to answer a Request For Information (RFI). As leadership you can ask the Cadre questions if they are in the form of an RFI. A couple of good questions are: 1) How many Cadets with rucksacks can fit on one silver side? 2) How many Cadets with rucksacks can fit on one bus? With this information, you can develop a plan and disseminate the plan to each squad by informing them what vehicle will transport them to the training area. Finally, think contingencies!

**Leadership Insight 17:** Cadets that ride on the bus sit two to a seat and have their rucksacks in their laps during travel. Splitting the platoon leader and the platoon sergeant between two vehicles is a great technique for maintaining command and control (C2). Don't forget, the platoon sergeant is in charge of accountability, which creates a major problem for most Cadets. During movement, the platoon leader should think ahead and be prepared for the next event (e.g. telling the PSG where to stage personnel and equipment, addressing the Cadre for specific instructions and risk mitigation).

**COLD HARD FACT 13:** Failure to get an accurate count of platoon personnel, weapons and equipment when loading and unloading vehicles will result in an emotional event if a Cadet is left behind or the movement is delayed. This is a prime time for Cadre Override!

## Daily Operations Order (OPORD) Process

After Cadets complete in processing and several other administrative actions, the regiment will begin to issue out a daily OPORD to disseminate information pertaining to the next 24 hours, usually around D3. The OPORD process is the most common platform for delivering information in the Army and is critical to ensure higher headquarters' operations are synchronized at all levels. While at LDAC, the regiment will exercise the same OPORD process to inform Cadets of the next day's operations. As stated before, the regiment will issue their OPORD at approximately 1800 or thirty minutes after the evening meal, therefore, always put the upcoming leadership ahead of the line for chow. Cadets must take good notes during this process because the RTO and Regimental SGM will ask specifics from the previous day information.

The *Regimental OPORD* briefing will require four days of company level leadership to be present. The outgoing leadership will brief an After Actions Review (AAR) for the previous day's activities, the incoming leadership will be issued a FRAGO, a WARNO will be issued to the leadership that will take charge in two days and an OPORD will be issued to the leadership that will take charge the following day. The intent is that company level leadership is exposed to the OPORD process and receives information for operations. The order of briefing sequence is Outgoing (AAR), Incoming (FRAGO), 2 days out (WARNO) and 1 day out (OPORD). Each leadership will be excused after their portion has been briefed. Each regiment will conduct their OPORD briefing with a different flavor but for the most part, they will all deliver the appropriate information. Figure 16 shows a variation of a Regimental OPORD Briefing Set-up.

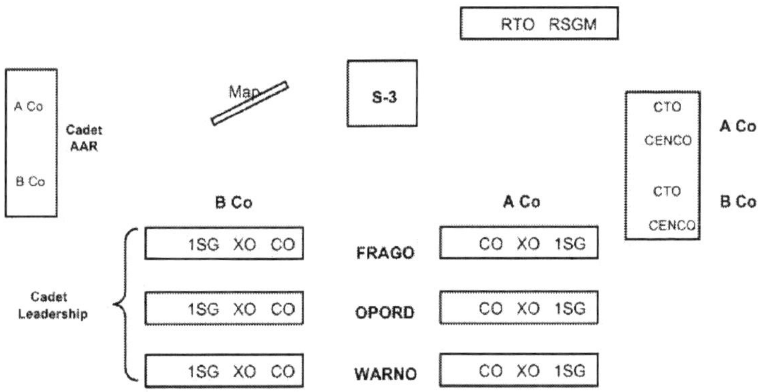


**Figure 15: Regimental OPORP Set-Up**

**Leadership Insight 18:** Company level leadership evaluations begin when you receive the WARNO 2 days out. I suggest taking notes! Immediately after receiving the Regimental FRAGO, the clock begins to tick so Cadets must adhere to the one-third / two-thirds rule to afford subordinate units the time to develop their Operation Orders. (e.g. if you receive your FRAGO at 1900 and lights out is 2200, you have approximately 1 hour to finish planning and disseminate the company OPORD to the Platoon Leaders; most of this OPORD should be completed before receiving the FRAGO.)

**COLD HARD FACT 14:** If you do not follow the one-third / two-thirds time rule and the 5 paragraph OPORD format, you will receive an "N" in the Domain Knowledge (DK).

Immediately after receiving the Regimental FRAGO, the Company Cadet Leadership is under the gun to issue a *Company OPORD* to the platoon leadership. If the Cadet is squared away, the majority of the OPORD will be completed before hand and you will have issued the platoon leadership a drop-dead time to receive the company OPORD. The main idea behind the one-third / two-thirds time rule is that higher headquarters issues an OPORD in a timely manner to allow the lower elements optimal amount of time to develop and issue their OPORD. This especially holds true for the company and platoon levels. Don't rush but don't take more time than needed to issue the OPORD. If the OPORD can be issued earlier than the one-third of the time requirement, then do so! The flip side to issuing a quick OPORD is to ensure that you have planned properly so that it does not affect your leadership evaluation. A poorly developed OPORD delivered early may not allow for good execution the following day. On the other hand, an over planned OPORD that exceeds the time limit screws your peers at the platoon and squad level. It's a juggling act of glass balls so don't drop one. The key to adhering to the one-third / two-thirds rule is task organization so that tasks can be divided and conquered. The Company XO and 1SG are there to assist the CO in executing necessary tasks. The Company XO should conduct a recon of the training area and route to the training area after receiving the Regiment WARNO. The XO should request transportation, weapons draw and turn-in, write paragraph V and ensure the Cadre is aware of any special requirements. The 1SG should determine the specified task, construct a timeline, set the uniform for the operation, coordinate for supplies and the dining facility and write paragraph IV of the OPORD. If all three of you are working together, the one-third / two-thirds rule will be a breeze. However, there will be times when operations don't go as planned and you may not receive the regimental OPORD until 2100. If this is the case, you have approximately 20 minutes to get the information out to the platoons because the 2200 lights out and the one-third / two-thirds rule NEVER EVER changes...EVER! My advice is to be proactive and stay alert for the leadership rotation schedule.

The *Platoon OPORD* process is exactly the same as the company OPORD process, except Cadets will have less time for planning and there are only two Cadets in leadership to develop the order. After receiving the company OPORD from the commander, your platoon evaluation begins. There are several key similarities that exist at all levels when preparing OPORDs. The first is the adherence to the one-third / two-thirds rule. It doesn't change! For example, if you receive the company OPORD at

2000, the platoon OPORD must be given to the squad leaders no later than (NLT) 2040 hours. Second, in order to adhere to the one-third / two-thirds rule, you must divide and conquer the necessary tasks by leveraging the Platoon Sergeant (PSG) to perform certain tasks. At a minimum, the PSG assigns specified tasks to each squad, starts necessary movements under TLPs, constructs a timeline, determines the uniform for the operation, coordinates for platoon supplies and writes paragraph IV and V of the platoon order. This allows the Platoon Leader to concentrate on the execution paragraph (Paragraph III). The third similarity is the location where the order is disseminated to your subordinate leaders. Cadets must select a location that is not in a high traffic area, has sufficient light and is a comfortable setting to avoid distractions.

The *Squad OPORD* mirrors the Platoon OPORD except on a smaller scale. Additionally, this is the lowest level in which an OPORD is graded by an evaluator. The squad OPORD must adhere to the one-third / two-thirds rule even though your team leaders do not normally give an OPORD. For example if you receive the squad OPORD at 2040, the squad OPORD must be given NLT 2103 to be technically correct but 2105 should be just fine.

Finally, there are several points that must be stressed in order for Cadets to succeed at any level during the TLP process. The Troop Leading Procedures (TLP) process measures several different Leadership Performance Indicators (LPI). These LPIs include but are not limited to Domain Knowledge (DK), Innovation (IN), Military Bering (MB), and Develops Others (DO). The IN and DK LPIs assist in evaluating on how you accomplish the mission. The DK LPI aslo evaluates if you use the 5-paragraph OPORD format and one-third / two-thirds rule. The Development Others (DO) evaluates how you empower your subordinate leaders and the Military Bearing (MB) LPI evaluates the Cadet's command voice and command presence while delivering the order.

**Leadership Insight 19:** Follow *"The Knuckle Dragger"* Approach. A common term in the Army is "Knuckle Dragger." Knuckle Dragger refers to individuals who don't have a large frontal lobe but can carry a hundred pound rucksack like there is no tomorrow. Knuckle Draggers do not have time for elaborate plans or multiple moving parts. They want to see pretty pictures that explain what is expected of them without a lot of thought. If you incorporate some key components in the OPORD process, you will follow the "Knuckle Dragger" Approach. The key components to receiving a good evaluation during the TLP process are: 1) Format – The use of the 5 paragraph OPORD format is a must so Cadets can issue a logical and natural flow of information to their subordinate leaders. The use of this format is the Army's way to communicate to lower echelons and it is very effective. 2) Graphics – Graphics assist the Cadet in disseminating information. Studies show that most individuals are visual learners; graphics are visual aids that allow Cadets to learn the order visually. At a minimum, use a map or draw a sketch to show the overall concept of the mission, write down the specified tasks and assign a squad to be responsible for that task, write out a timeline and construct a uniform and rucksack packing list. These graphics are best displayed on butcher block; an inherent task is to ask for butcher block if there is not one provided. Finally, don't forget to write these tidbits inside the Cadet TACSOP before leaving for LDAC. 3) One-third / two-thirds Rule – Following a time standard is a must or else Cadets at the lower levels will not be afforded the opportunity to develop a reasonable plan. Failure to follow the one-third / two-thirds rule will trickle down the chain and cause an emotional event for the squad leaders.

**COLD HARD FACT 15:** Failure to adhere to the one-third / two-thirds rule, failure to use the 5 paragraph OPORD format, failure to use graphics or deliver an unobserved OPORD during TLP will result in the Cadet receiving an "N" in the Domain Knowledge (DK) indicator at a minimum. Failure to have a Cadre present while delivering your OPORD at any level results in an unobserved OPORD and a BIG FAT "N" in those multiple indicators.

**Garrison Training Events**

During LDAC, your platoon will transform from individuals into a tight knit team at the platoon level and even more so at the squad level. There are many different training events that are either individual or collective as a group. These training events may or may not be conducted at LDAC every year but Cadets will experience these or similar training events to serve as initial exposure to a prolonged duration of rigorous training. Training events consists of Land Navigation, CWST, Basic Rifle Marksmanship, US Weapons (Squad Automatic Weapon [SAW] and the M60 machine gun). The training is used as a familiarization because both weapons will be used in the field portion of LDAC. There is a day for an Obstacle Course, Confidence Course, Leaders Reaction Course and Rappel Tower. The CPT Michael Tarlavsky Hand Grenade Assault Course is an absolutely fun day that affords Cadets the opportunity to practice throwing grenades, identify the type of grenades, experience a grenade assault course and culminate with throwing a live hand grenade. Another fun day is the Individual Tactical Training (ITT) that covers individual battlefield skills, combat movement techniques and procedures that build toward squad and platoon battle drills. Fire Support training teaches Cadets the importance and lethality of artillery fire on the battlefield, employment of indirect fires, and gives them the opportunity to employ these techniques during field operations. A not-so-fun day is the Nuclear, Biological, Chemical (NBC) training that provides basic soldier skills that cadets must master to meet pre-commissioning requirements. Cadets learn characteristics, maintenance and the employment of NBC equipment through an exercise called Cobalt Hill that incorporates mask confidence exercises to develop confidence in soldier's equipment. A fun game is to challenge the Cadre to a push-up contest in the Gas Chamber if you think you got what it takes to win. Finally, after you return from the field training of LDAC, your platoon gets to enjoy a day at the lake conducting water confidence training in Zodiac boats, executing the log walk and rope drop followed by negotiating a seventy-five foot slide for life.

**Leadership Insight 20:** Some training events such as the land navigation training are a must pass event. The confidence training is a must attempt event. Failure to attempt an obstacle will result in a counseling statement for "Failure to Train."

**Leadership Insight 21:** During Garrison evaluations, eleven different Leadership Performance Indicators (LPI) are almost always seen. (Refer to Table 1.)

**COLD HARD FACT 16:** A "Failure to Train" counseling statement results in a removal from Warrior Forge without LDAC credit.

**Night Hawk** is a committee that teaches patrol base classes and evaluates Cadets during patrol base activities after conducting SSTX lanes. Although LDAC has designated this as a committee in the past, it can be done by any ROTC Instructor and may be the wave of the future since ROTC is constantly losing resources to the needs of the Army. Not all Cadets will receive two evaluations during Garrison phase at LDAC because there are not enough days for the amount of people attending LDAC. Night Hawk or FOB activities are used to provide a second Garrison Evaluation for those Cadets who still need an evaluation. If a Cadet already has two evaluations, the evaluation counts as a positive or negative spot report. Cadets depart

Garrison and arrive at a field reception area for a briefing. Immediately following the reception, Cadets receive classes in Cultural Awareness and Patrol Base Activities.

Cultural Awareness gives Cadets the insights to how other cultures may not accept the U.S. policies or procedures. Both females and males can have a very difficult time leading when these variables are implemented into the SSTX lane. Females can expect the indigenous forces not to talk to them or respect them as leaders. If this happens, it is up to the Cadet to work through this issue and still lead. It may require using the subordinate leaders to gain rapport and then ease into the situation; however, always remember, no matter the gender, all Cadets can expect to be placed into uncomfortable situations during their leadership.

Patrol Base classes are given to Cadets at the platoon level for the remainder of the day. Some Cadets will receive an evaluation that can count toward their second Garrison Evaluation on their ability to effectively set-up a patrol base and perform patrol base activities. I can't stress this point enough but if a Cadet has already received two Garrison Evaluations, this evaluation will count toward a positive or negative spot report which could give you that opportunity of receiving an overall "E" Rating. This is a great chance to gain more LPI Es. Night Hawk is a very beneficial tool at LDAC and really separates the top Cadets from their peers. In 2009, the Night Hawk Committee was not utilized.

# 8 FLRC (The Tough "E")

The Field Leadership Reaction Course (FLRC) is designed to develop and evaluate leadership while building teamwork early in the LDAC cycle of training. This may very well be your first leadership position at LDAC and it comes at you fast and furious. This leadership opportunity stresses problem solving, critical thinking and communication skills. Cadets will conduct the FLRC as a squad that will rotate through different stations along the course that are designed to last approximately 45 minutes and give early feedback on a Cadet's leadership strengths, weaknesses, styles and techniques.

Most Cadets attending LDAC believe the FLRC Blue Card Leadership evaluation is very ambiguous and the 2LT graders are out to get them and protect the "Butter Bar" rank of 2LT. History has shown that the majority of the Cadets will receive an "S" rating on their FLRC evaluation unless they are squared away and treat the FLRC event as a mission by using the Troop Leading Procedures to their advantage. Historically, 20 – 25 % of the regiment receives an "E" Rating.

You will conduct FLRC during Garrison Operations and it

will be counted as a separate evaluation so *Troop Leading Procedures are expected.* Once you arrive at the FLRC site, you will be divided into squads and placed into a rotation on the course so every Cadet receives a Squad Leader evaluation or Leadership Assessment Report (Blue Card). Your squad will arrive at a station on the FLRC course and an Evaluator will call out a name and place that Cadet in charge for that station. Once assigned as the leader for that station, the Cadet receives initial instructions followed by a more detailed mission brief. After receiving the mission, it is absolutely crucial the leader follows TLPs. A key thing to remember is Cadets are being evaluated on leadership and not the ability of their squads to complete the obstacle. Key tasks that are being evaluated are the ability to utilize Troop Leading Procedures, conceptualize a quick but solid plan, brief with confidence, utilize a modified OPORD format, employ graphics such as a notebook sketch or an improvised terrain model, use team leader recommendations, leader location to control the mission and the ability to empower team leaders. Remember to be positive, maintain positive control and lead first and complete the task second. Below are some pointers and how to implement the eight TLPs.

*Receive the Mission*

Cadets receive a scenario briefing with limitations instead of a traditional 5-paragraph OPORD for the FLRC event. Remember, leader actions are being evaluated while receiving the mission brief so be sure to listen, take notes and ask questions to confirm the mission. There will be safety factors that must be observed and limitations that will deduct time during the event causing leadership challenges to evaluate how the Cadet in charge handles the situation. If you fail to maintain safety, it will be a major demerit on the evaluation.

*Issue a Warning Order*

Issue a Warning Order immediately after receiving the mission brief in order to allow subordinate leaders to quickly brief personnel in the squad. Be brief when giving this order; it should not take a long time and should only contain minimum information.

*Make a Tentative Plan*

Develop a quick plan at the location where you receive your mission brief before returning to the squad area. If you do not have any idea of how to negotiate the course, ask for input from your team leaders but don't let them take over your leadership evaluation or it will become an emotional event. Team leaders can advise squad leaders of a plan but the leader must make the decision to implement the plan; this allows the leader to maintain control of the situation and get subordinate leaders involved.

*Initiate Movement*

Start movement before leaving to receive the mission brief by giving specific orders to team leaders. Example of specific orders can be to ensure all personnel have water, conduct PCC/PCIs of personal equipment and un-blouse their ACU pant legs.

*Conduct a Recon*

Conduct a recon at the same time when receiving the mission brief by drawing a quick sketch and take it back to the squad area or by asking the Evaluator to allow the squad to come forward to view the obstacle. The latter of the two is the best option because everyone will have a chance to see what the squad must negotiate. If the squad is brought forward during the recon, use this same area to give out your modified OPORD.

*Complete the Plan*

This must be completed very quickly if you return to the squad area! Issue a modified OPORD that includes task organization (near and far side elements), delegate tasks to each fire team to include equipment preparation, advise personnel on station limitations and safety considerations and use a sketch or draw a graphic representation in the sand using sticks and rocks to brief the plan. Either way, this becomes a sand table for briefing, which is a major plus for the evaluation.

*Issue the Plan*

Issue the plan **NO LATER THAN 15 minutes** after receiving the mission! Remember there is only 45 minutes to complete and execute the plan so lollygagging is not necessary. Therefore, effective use of your time is necessary to gain efficiency and adhere to the one third / two thirds rule! This is very obtainable as long as you task organize and don't second-guess your plan. Things to remember when briefing are to make clear and concise explanations of tasks to be accomplished, ensure safety factors are clearly expressed, emphasize team effort and answer team members questions.

*Supervise*

As a leader, you must develop the plan, issue out the orders and then supervise the actions of your squad. During the execution of the obstacle, supervisory skills are evaluated, including how well the leader maintains control of the squad. Other factors and questions asked by the Evaluator are:

- *Did a team member take charge of the team?*
- *Did the leader keep composure during the entire problem?*
- *Did the Cadet emphasize safety to the squad during the problem?*
- *Did the Squad Leader modify the solution during progression based on team input?*

During each event, problems or issues may arise that could hinder the completion of the obstacle using the current plan. If this occurs, ask for suggestions from team members and continue to motivate them to be problem solvers. Remember the Golden Diamond because this is a prime opportunity to implement the cycle of assess the situation, introduce a new plan (if needed), make a decision (if needed) and execute for that hard to get "E."

The Golden Diamonds shown in figure 17 equate to the process of using four Leadership Performance Indicators (LPI) that demonstrate a Cadet's ability to understand and implement the complete decision-making cycle while working through sit

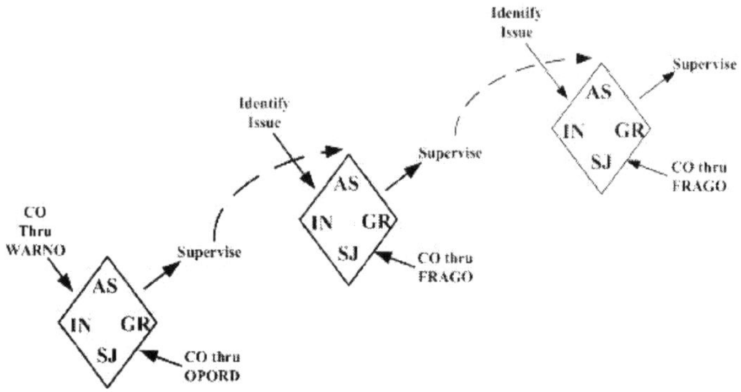


**Figure 17: Golden Diamond Process**

uations or problems. The first cycle of the Golden Diamond Process starts with assessing (AS) the situation, Innovating (IN) a course of action, if necessary, make a decision using sound judgment (SJ) and finally executing the plan by issuing a Warno OPORD (OPORD) to get results (GR). During this process, the WO and OPORD measure the communication (CO) behavior. After issuing the OPORD, the leader uses step eight of the Troop Leading Procedures to supervise the execution of mission and begins the cycle over again by assessing how the plan is working. If a problem is identified after assessing the plan, the leader innovates how to adjust the plan, makes a decision if necessary and issues a FRAGO; the process continues again if necessary. The Golden Diamond is what separates great leaders from ordinary leaders! A Cadet who can implement and utilize this process is more likely to be in the Platoon's Top 5, be a front runner for the #1 in the Regiment and be very successful in any career field.

# 9 Assembly Area and Forward Operational Base

Prior to 2002 very few soldiers knew about the concept of a Forward Operational Base (FOB). The Forward Operational Base concept was part of a series of Operational Bases from the Special Forces (SF) Company to the SF Group (brigade) level that allowed the SF group commander to exercise Command and Control (C2) at the lowest levels. The main idea behind the FOB is that the FOB combines the functions of a command post, unit trains and a staging area into a single entity. A FOB by doctrine supports a battalion and an Advance Operational Base (AOB) supports a company.

By military definition a FOB is a command, control, and support base established and operated by a battalion or higher using organic and attached resources. The task organization and functions vary with the factors of METT-TC and OAKOC or OCOKA. The FOB may be located with its higher headquarters or many miles away from it and is responsible for preparing, deploying, controlling, sustaining and tracking its subordinated units. Main Operating Bases provide backup support and traditionally support FOBs that are used to support tactical operations. Finally, a FOB may be used for an extended period of time and may or may not contain an airfield, hospital or other key facilities.

**LDAC FOB Overview**

At LDAC, Cadets may receive one of their Garrison Operations while occupying a FOB or Assembly Area (AA) in a field environment. Many Cadets look at this leadership as a challenging event; however, in reality it is a lot easier than a Garrison evaluation because the ever-popular training distracters such as the Post Exchange (PX) or phones don't exist in the FOB. In the same respect it is VERY difficult to get an "E" Rating because there are only a few activities that must be accomplished in the morning before leaving for training.While at the FOB, Cadets will be responsible for many different tasks, including preparing for combat operations and conducting training; however, Cadets will not be required to construct any major modifications. The FOB at LDAC is of a small scope compared to what was mentioned above; however, it does meet the definition of a FOB and allows for Cadets to refit and prepare for field operations. Additionally, it gives Cadets attending LDAC an opportunity to experience FOB operations during their movement into Patrolling Situational Training Exercises (PSTX). Prior to arriving at the FOB, your regiment will perform a link-up with the Palomians. After your link-up, the Palomians will lead you to the FOB by foot, or bus the regiment to a dismount point depending on LDAC scheduling. If a foot movement is conducted, it will be approximately 10 kilometers simulating your entry into the country of Paloma. If the regiment is transported to the FOB, the regiment conducts the 10 kilometer foot march after PSTX; either way, Cadets will do a 10 kilometer foot march! Upon your arrival into Paloma, your squad will occupy a position inside the FOB where you will be allowed to set up an area to refit before moving into PSTX. During this time, Cadets receive their duffle bags with fresh uniforms, socks and other necessary items for PSTX. After you refit, your squad is married up with another squad to form a Patrol Element. The Patrol Element then receives classes for the remainder of the day in preparation for PSTX. Every Cadet will get a leadership during PSTX, which is usually the final leadership evaluation. After classes, patrol elements are turned over to your Patrolling TACs for FOB security. Once the FOB is occupied, be prepared to be probed by the enemy and visited by protestors. The FOB leadership will be evaluated on how they handle these situations. During the late afternoon or early evening, each Patrol Element will depart the FOB and conduct a foot movement to a Patrol Base which will begin PSTX giving Cadets their final opportunity to get an "E" Rating. The main difference between SSTX and PSTX is the amount of time for planning, the amount of people in the element, the distances you travel between missions and the additional leadership positions for evaluation. Some cool things that you will experience are a helicopter ride (resource dependent), constructing a poncho hooch, massive size anthills and some very hungry and BAD ASS mosquitoes. Make sure you prepare accordingly; bring Deep Woods Off and a mosquito net.

One final note is this will not be the last time you experience FOB Operations in your career because they will be seen in future deployments into Iraq and Afghanistan.

**FOB leadership** can go in many directions especially if Cadets have not received any guidance toward this process. The best procedures in the absence of guidance are to refer to Field Manuals or Standard Operating Procedures. Since most Cadets are not familiar with either one of these aids, execute the following tasks and life should be okay.

- Identify any minor/major deficiencies in the obstacle effects around your perimeter that need replacing/reinforcing and make a list of these as well as a request for supplies to fix the problem (you likely won't get the supplies but this action demonstrates conceptual thinking). Ensure to cross level ammunition for the FOB security force and adhere to the Paloman rules for the FOB because of environmentally sensitive areas.
- Develop a detailed terrain model in a central location so leaders can brief the security plan.
- Identify Traffic Control Points (TCPs) and over watch positions guarding your perimeter. Identify areas of responsibility to include individual sectors of fire and fields of fire. This will assist leaders in developing a security plan for the FOB which becomes very important when security is reduced to 50% or less. Identify what over watch positions

must be manned in order to have 360 degree security so that a rotation can be accomplished. This is a very sensitive area and Cadets must understand the Rules of Engagement (ROE). If a Cadet shoots a Paloman civilian it is a negative mark against the leadership and your stay in Paloma can be very difficult. Cadets must understand who the friendly Palomans are and who are enemy combatants so proper actions can be taken.

- Implement the 7 steps of Engagement Area Development to plan when constructing FOB Defense/Security. The first three steps are considered mission planning and the final steps are considered execution of the plan.
- **STEP 1.** Identify all likely enemy avenues of approach. Accomplish this task by looking at the FOB from the enemy's perspective based on METT-TC and OAKOC. Determine which routes the enemy would attack the FOB. Determine which avenues personnel must protect and determine which avenues can be protected using obstacles.
- **STEP 2.** Determine likely enemy schemes of maneuver. This is the step where you take METT-TC, OCOKA, enemy's capabilities and TTPs and the enemy commander's point of view into consideration when determining how the enemy will attack the FOB. This should result in the enemy's Most Probably Course of Action (MPCOA) and a Most Dangerous Course of Action (MDCOA). For example, "Will the enemy attack the FOB using small arms fire, mortars,

local populace or a diversion?" Once the enemy's scheme of maneuver is discovered, determine a course of action on how to defend the FOB against the MPCOA and MDCOA. Hint: Artillery or Mortar Fire should be incorporated into your plan.

- **STEP 3.** Determine Where to Kill the Enemy. Using the enemy's most probable and most dangerous course of action; determine the primary and secondary Engagement Areas (EA) to defend the FOB. The next three steps are the actions on the objective for FOB Operations.
- **STEP 4.** Emplace Key Weapon Systems. Crew Served and Key Weapons Systems will be used to cover Engagement Areas. Weapon Systems may include M240B, M60, AT4, M249, M203 and Claymores. Emplace these key weapons in positions to cover your engagement areas effectively. You may have a tower or over watch position covering your primary EA where you want to place key weapons and identify a Primary Direction of Fire (PDF) and a secondary sector of fire; remember the Range Card for a Patrol Base? This is an excellent time to use a Range Card and will get you noticed in a positive way. The M203 is an excellent weapon to cover dead space and augment the crew served weapons. Target Reference Points (TRPs) can be used on key terrain features and EAs outside the FOB. The integration of all these weapon systems will be key to having an effective defense/security plan.

- **STEP 5.** Plan and Integrate Obstacles. Obstacles are used to channelize or slow down an enemy advancement. Although Vehicle Check Points (VECP) will have concertina and crude jersey barriers it is unlikely you will be given any other barrier materials (Class IV) to construct obstacles but plan for the materials and ask for the world to demonstrate conceptual thinking by providing solutions against a known enemy.
- **STEP 6.** Plan and Integrate Indirect Fires. Always make a plan to incorporate indirect fires into a defense plan since a FOB is considered a defensive operation. Assign Indirect Fire Targets to support key locations such as large open areas, road intersections, easily identifiable structures, etc. outside the FOB/AA. These Indirect Fire Targets should integrate with and compliment Obstacles and Key Weapon Systems in order to create a defense that integrates multiple assets.
- **STEP 7.** Rehearse. Always rehearse the plan no matter how minimum or how short on time. Rehearsals identify gaps in the plan and synchronize movement and responsibilities and should include occupation of the FOB, reinforcing obstacles, occupation of security positions and reacting to an enemy attack.

Finally FOBs are the future of our conventional forces mainly due to the fact we are fighting a fourth and fifth generation of warfare. Special Operations have always fought this Asymmetrical Threat since their existence, which required communications and support units be pushed closer to the combat units. After returning from LDAC last year, a Cadet asked the question, "How are FOBs being used in Iraq and Afghanistan?" From my experience and perspective, FOBs are currently being used as a soldier's "home away from home" to allow soldiers the opportunity to take a knee or gain relief from constant danger, minimize the draining effects of hunger and fatigue and reduce the isolation of combat. As a result, many of the factors of psychological stress typically experienced by soldiers in combat are greatly reduced because they have some type of facilities to sustain them for a long period of time. These services usually consist of phones, Internet, Shopette, laundry services, Green Bean Coffee Stand (Well Valued) and local national shops to allow the soldier to relate to life back in the states while having a place they can call home.

# 10 Platoon and Squad
## Situation Training Exercises

The Purpose of SSTX and PSTX Lanes is to evaluate Cadets on their potential to become a future Army Officer in a field environment. **Squad STX** is a two-phase even used to evaluate basic leadership using squad battle drills and collective tasks. The first day is used for classes, the next two days are used to evaluate Cadets' squad tactics and the remaining two days are used to evaluate leadership by adding variables to the tactical scenarios resulting in an increase of your stress level. SSTX lane duration is two hours and can encompass missions such as Small Unit Tactics, dealing with the press, civilians on the battlefield or a combination of all the above (called variables). The mission can be changed during the execution phase (known as a FRAGO) to evaluate the Cadet's ability to be flexible and adapt to change. Each Cadet receives two formal evaluations of his or her performance as a squad leader that requires knowledge of land navigation, terrain association, weapons systems, and all individual training previously presented in this book. **Patrolling STX** is an event that employs two squads of personnel to form a section allowing Cadets to lead a larger element for a longer duration. The first day is a training day on which Cadre teach the fundamentals of patrolling missions, followed by two days of leadership evaluations as a Patrol Leader and Assistant Patrol Leader using tactical scenarios. Patrolling STX is conducted immediately following SSTX in order to encompass all previous training and increase a Cadet's level of performance while leading a larger unit. Cadets receive one formal evaluation during this phase and learn about basic air assault operations. One final note is that every year the planners at Warrior Forge attempt to make the training better than last year and incorporate lessons learned from the previous year. It is possible that the length of these two training events may be shortened or extended in the future.

Both the PSTX and SSTX have specific procedures that can be followed to reduce the stress and difficulty of field operations. These procedures must be practiced and tailored to each Cadet in order to become proficient and comfortable leading during this type of environment. The remainder of this chapter discusses major events that must take place during SSTX and PSTX lanes to include how to determine and translate pertinent information.

The **Take Charge Minute (TCM)** is the most important event that takes place prior to a Cadet conducting a SSTX or a PSTX lane evaluation and sets the tone of your leadership. Before a Cadet conducts a SSTX lane, it is beneficial for the squad to create SOPs for operations so the leader can reiterate them to the squad; this includes on campus and at LDAC. Although the TCM is not the Warning Order, it initiates movement of the squad and allows squad members to support you and be a team player.

***TCM Example:***

*Hello, my name is CDT Excellence from the University of Leadership. I will be your Squad Leader for the upcoming PSTX or SSTX lane. While I receive the OPORD from higher, I need the following to take place:*

***Alpha Team:***

- *Divide ammunition between elements for a 60/40 split; support will have 60% to facilitate the mission.*
- *Begin terrain model construction; one terrain model for movement and overall concept of the mission and one for Actions on the Objective.*
- *Conduct PCIs for your team and ensure they have all mission essential equipment.*
- *As per SOP, your jobs will be Primary Aid and Litter, Civilian on the Battlefield Team (CB Team), Translator Team (usually CB team), Secondary Demo, Secondary EPW, and Secondary Breaching/ Bunker team.*
- *Conduct team rehearsals.*

***Bravo Team:***

- *Conduct water resupply.*
- *Conduct PCIs for your team and ensure we have all mission essential equipment.*
- *As per SOP your jobs will be Primary EPW and search teams, Demo, Breaching/ Bunker Team, Secondary Aid and Litter, Secondary CB Team, and Secondary Translator Team (if possible).*
- *Conduct team rehearsals.*

*You will have 10 minutes while I am receiving the OPORD to complete the following tasks. I want all these tasks complete in 5 min shifts. Alpha will pull security first while Bravo completes their tasks, then Bravo will pick up 360º security and Alpha will complete their tasks. If you complete your tasks before your 5 min is up, begin rehearsing special teams placing priority on Aid and Litter and EPW. Once I return, team leaders come back to the center for WARNO.*

**Receiving the OPORD** from Evaluators during PSTX or SSTX Lanes can be an emotional event if you are not prepared to receive the information. You must have a game plan to receive the information because it will come at you fast and furious; but on a good note, not every bit of information that is given to you is needed. Therefore it is very crucial to understand what is important and what is not. If you have a game plan and understand what you must obtain from the higher OPORD, you will be able to focus your attention to the pertinent and necessary information. Once the leader is determined for the upcoming SSTX, there is approximately 5 minutes for the Squad Leader to issue the TCM to the squad. Afterwards, the leader will follow the TAC to a briefing area to receive the OPORD from a sand table that represents the SSTX lane to include the lane direction; it reminds me of a golf course individual hole outline. Below are a few vital things to remember during the OPORD process to allow maximum use of time for creating the squad level OPORD much faster:

Make a skeleton OPORD in your notebook to fill in the blanks quickly. Although you are currently not allowed to have laminated OPORD templates, you can prepare one by hand in your field notebook before taking charge of the squad. Ensure that you have enough room to annotate information for the OPORD so that you do not become confused after receiving the OPORD. I recommend using one page of your notebook for the Situation and Mission paragraph; one to two pages for Execution; one page for Service & Support and Command & Signal.

Assign the RTO specific items to write down during the OPORD from the TAC (e.g. friendly units, mission statement, call signs and grid coordinates) so you can focus on Para III. Once the information is received, there are two options. 1.) Release the RTO after paragraph III to start constructing the sand table or 2.) they can remain for the entire OPORD. The main reason for sending the RTO back to the squad patrol base is to inform your squad of the upcoming mission and to finish the terrain model. It is your call! Remember, you are in charge!

Knowing what pertinent and necessary information to listen for in the Higher OPORD from the SSTX TAC is half the battle because much of the information from the next higher level OPORD is not needed for the Squad OPORD. Of course, the information varies based on the paragraph but there is an explanation below to make it a bit simpler:

Cadets receive the OPORD in a 5-paragraph format with each paragraph containing pertinent information that you must write down in order to prepare your order. Below is the minimum data that you must obtain from the Platoon OPORD so that you can develop your Squad OPORD. It is important to understand what information you and your squad need to know so that the Squad OPORD can be completed in a timely manner. Over the next section, we will discuss the 5-paragraphs of the platoon OPORD that you will receive from the TAC. Each paragraph is broken down to the pertinent information that you must obtain so that you can give a squad OPORD.

**Deriving and Translating Pertinent Information**

The next step is to write the Squad OPORD by deciphering and translating the pertinent information from the information that was just received. The *Higher OPORD Situation paragraph* contains useful information about the enemy and friendly units within the Operational Environment (OE) and assists in the development of what action to perform. Enemy pertinent information is the number of enemy personnel or the size of elements that are conducting operations, (i.e. squad size, 2 or 3 man elements), the type of weapons they use, what type of operations they are conducting and if they have reinforcement capability. Friendly information will be about the friendly unit's locations and their general mission but the only real information you must record in reference to friendly units is the friendly squad's location in reference to your squad so that you can conduct lateral coordination. This is also important because it will tell you who can provide security for you while you brief your OPORD. Remember you have to ask higher for another squad to conduct security for you while issuing your OPORD but don't get too complacent because once on active duty it is not that easy.

The *Higher OPORD Mission Statement* and other locations have valuable information that allows you to generate a proper Squad Mission Statement. The platoon mission statement answers the 5 Ws of Who, What, When, Where and Why for the platoon mission and contrary to popular belief, the platoon mission statement has very little information that will assist you in developing a squad mission statement. However, it does tell you what the platoon's mission is and the overall concept of what is going to be accomplished. For example, if the platoon's mission is to conduct a movement to contact, then you can bet that your squad mission will be to conduct a movement to contact within a specific area in support of the platoon.

The main location in paragraph III that Cadets should become familiar with is sub-paragraph 3b (Tasks to Maneuver Units) from higher's OPORD. Remember this when you copy information down from the TAC because it is very important to get the mission correct and realize that the maneuver discusses the nuts and bolts of dealing with the Concept of the Operation.

**Leadership Insight 22:** Most Cadets make the mistake of writing down the platoon mission statement instead of formulating a squad mission statement.

The *Higher OPORD Execution paragraph* has pertinent information that pertains to all squads within the platoon. Some of this information will be specific to your squad and some information will be for the entire platoon (Coordinating Instructions). You must be able to obtain the necessary information from paragraph III because it will affect your squad's mission. Below are the sub-paragraphs of paragraph III and what information you should obtain from each:

3a. Concept of the Operation sub-paragraph will give the generalities and overall concept of the operation.

1. The Maneuver section tells the specifics and phases of the operation and what each squad does to support the platoon mission during each of these phases.

2. The Fires section states which squad has priority of fires from artillery or mortars during the mission. Information from this section can assist in supporting the mission by establishing a known location (Target Reference Point) for quick delivery of artillery rounds to assist you in a withdrawal from the objective; an additional use can be used to initiate an attack.

3b. Task to Maneuver Units gives each squad its specific mission and task that must be accomplished to make the mission successful. Combine this information with the information gained from the platoon mission statement and the Concept of the Operation and you have all the information required to develop the squad's mission statement. Believe it or not, **IT'S THAT SIMPLE!**

3c. Task to Combat Support Units is rarely used but would address specific tasks to units such as Engineer or Military Police units supporting the operation.

3d. Coordinating Instructions is a catch all section that addresses any issue that is not specific to an individual element. The main information in this paragraph is the time line, Priority Information Requirements (PIR), Rules of Engagement (ROE), treatment of Civilians on the Battlefield and NBC threat.

The *Higher OPORD Service and Support paragraph* signifies a decision point for the leader because the Radio Transmitter Operator (RTO) may either stay to assist you in gathering the remainder of the information or the RTO can leave to take over construction of the sand tables and deliver important information such as the grid to the OBJ to the team leaders. This paragraph tells when the resupply will arrive, any additional equipment or ammunition you have such as machine guns, binoculars, or smoke. If no equipment is mentioned, ask for it! (e.g. Claymore for an ambush, AT-4 for Knock out a Bunker or Binos for a Recon Mission). Important grids are given out for the Company Casualty Collection Point (CCP), Enemy Prisoner of War (EPW) Collection Point and Company Assembly Area. In real world operations, these grids are crucial for the MEDEVAC plan, securing EPWs from the objective and where your higher commander Command Post is located. At SSTX, you must put the information out to team leaders because there is not enough time to move casualties back to the platoon CCP or transport EPWs to the EPW collection point. Rehearsals are necessary to be able to conduct this type of drill.

**Leadership Insight 23:** The majority of Cadets fail to plan for an EPW or CCP since they think the platoon CCP is the correct location because it is given to them; this is not the case. When producing the Squad OPORD, it is important to establish a CCP closer to the objective in case your squad receives casualties. If the squad takes casualties, they should be treated, taken to the ORP and then be evacuated to the platoon CCP.

The final paragraph is *Command and Signal* where Cadets receive location of key leaders, passwords, chain of command and call signs. You should write down all of this information because you will need it for battle drills and contacting higher when things don't go as planned. Additionally, you will need to know the call signs of your higher to request permission to cross the LD, coordinate for security and call the Forward Observer if you use artillery or mortars.

As stated above, there is a lot of information that must be

obtained from the higher OPORD before being able to have enough information to construct a squad OPORD. Since there is approximately 15 minutes to deliver your OPORD, sifting through the extraneous information and only writing down the pertinent information will save time by ensuring only necessary information that affects your squad is delivered to accomplish the mission.

The best way to **translate the pertinent information** from the Company OPORD to the Platoon OPORD or from the Platoon OPORD to the Squad OPORD is to be organized before receiving the information because it allows you to translate the information very quickly without having to rewrite the information upon returning to the Squad Assembly Area. A sample format is shown below for each of the five paragraphs.

The *Situation paragraph* has very little translation but contains important information about the enemy and friendly units. The most significant thing you should take away from this paragraph is to ensure you answer the "So What." The "So What" in this case is to take the enemy information given to you from the higher OPORD and determine the Enemy's Most Probable and Most Dangerous Course of Action. As for the friendly situation, it is a straightforward approach by repeating what was given to you in the Platoon OPORD. Organize your notebook as follows:

**Enemy Composition**
What the enemy is composed of (e.g. uniform, number of personnel, and type of weapons or equipment).

**Enemy Disposition**
The area the enemy is operating (e.g. AO SLAP), should include a grid coordinate or an area. Normally this is where your objective will be located.

**Enemy Capabilities**
What type of operations the enemy is conducting (e.g. enemy forces are operating in 2 to 3 man elements conducting small unit tactics; raids, supply depot operations, recon, etc). Can they be reinforced with larger units; if so how (vehicle or on foot) – This will assist you in determining how long your element can remain on the objective.

**Leadership Insight 24:** Information gained from the enemy portion of paragraph I should enable leaders to formulate the Enemy Most Probable (EMP) Course of Action (COA) and the Enemy Most Dangerous (EMD) COA. Keep it straight and to the point!

**EMPCOA** – Tells the squad what the enemy is currently doing; will the enemy stay and fight or will they run? (e.g. the enemy will continue to conduct recon and intelligence operations in AO SLAP; if detected they will stay and fight.)

**EMDCOA** – Tells the squad what the most dangerous action the enemy can do that would affect your operation (e.g. if attacked, enemy forces will call for a platoon size reinforcement that can be on the objective within 10 minutes).

The Mission paragraph translation can be simple but confusing; however, it is very important to understand because it shows that you have good technical skills, communication and knowledge. Since the mission statement must include the 5 Ws: Who, What, Where, When and Why, a good technique to write your mission statement is to write down this format in your notebook prior to receiving the OPORD (shown below). This allows you to be ahead of the game when the platoon order comes at you fast and furious. The words in **bold** are constants and underlined words are mission dependent. An example is shown to illustrate what information you need from the platoon's order to formulate a squad mission statement. The mission dependent words are underlined.

**EXAMPLE:**

*TAC Reads:*
*Platoon Mission Statement and Tasks to Maneuver Units*

*Mission: A Co, 1st **PLT** conducts area ambush **NLT** (H+80) 15 May 2008 to destroy enemy patrols vic. ET 129459 in order to **(IOT)** prevent the enemy from locating avenues of approach into BP YANKEE.*

*Tasks to Maneuver Units: 1st SQD, the main effort, cross the LD (H+45), move along Route Blue and **conducts** a point ambush at ET 123456, set **NLT** (H+80)*

Once the TAC completes the OPORD, he or she will state the current time (e.g.1340).

**Leadership Insight 25:** The mission statement must include the 5 Ws: Who, What, When, Where and Why. Your mission statement is a combination of the information from the Platoon's Mission Stement and the Platoon's Task to Maneuver Units (this pertinent information is underlined).

*Who: 1st Sqd*

*What:* ***conducts*** *point ambush*

*When:* ***NLT*** *1500 15 May 2008.*

*Where:* ***IVO*** *ET 123456*

*Why:* ***IOT*** *prevent enemy forces from locating avenues of approach into BP Yankee.*

*1st SQD* ***conducts*** *point ambush* ***NLT*** *1500 15 May 2008* ***IVO*** *(in vicinity of) ET 123456* ***IOT*** *(in order to) prevent the enemy from locating avenues of approach into BP YANKEE.*

**Leadership Insight 26:** Although the OPORD states cross the LD NLT H+45 and conduct ambush NLT H+80, you must convert this to actual time. You accomplish this by listening to the time the TAC gives you (e.g. "The time now is 13:40, what are your questions?") 13:40 becomes your base time for everything for this operation. Cross LD at 1425 = 13:40 + 45 min; Conduct Ambush at 15:00 =13:40 + 80 min.

When translating the Platoon Execution paragraph into a Squad OPORD, the Concept of the Operation should tell an overview of the operation that is to be conducted. The pertinent information received from the platoon OPORD allows you to create a plan because this is the meat and potatoes of the mission and must be accomplished quickly and related to the terrain model. This paragraph has four sub-paragraphs that are discussed below and should contain pertinent information that you received from the platoon OPORD and the remainder comes from your imagination and how you will accomplish your mission.

**EXAMPLE:**

This is a 5-phase operation consisting of: pre-deployment, movement, employment, return movement and post deployment.

**Leadership Insight 27:** Divide the maneuver paragraph into these 5-phases.

***Phase 1 – Pre-Deployment or Pre-Movement***

Phase 1 includes all information concerning your squad before they cross the Line of Departure (LD) and is referred to as Assembly Area (AA) activities or a Patrol Base (PB) activities. Activities within this phase are Troop Leading Procedures, conduct rehearsals, water resupply, food pick up, ammo pick up or redistribution. The allotted time during SSTX is 45 minutes, which starts after you receive the platoon OPORD. That is why the platoon OPORD states LD (H + 45). The time line breakdown is:

15 minutes to prepare you OPORD

15 minutes to give the OPORD

**15 minutes for rehearsals**

45 minutes Total Cross LD

### *Phase 2 – Deployment or Movement*

This section addresses how the squad will move from the AA to the Objective. The basic information that should be included is the primary and secondary routes, distance, direction, movement type (foot, vehicle, or aircraft), movement technique (traveling, traveling overwatch, bounding overwatch), and the order of movement (OOM). The OOM should be in accordance with the team that was assigned primary compass and pace.

**EXAMPLE**

*1st squad will be traveling by foot on a 270° azimuth for 400 meters along the primary route (Route Blue). A-Team will be 1st in OOM using the traveling overwatch movement technique. A-Team will be responsible for establishing enroute rally points every 200 meters.*

### *Phase 3 – Employment*

Phase 3 is the meat and potatoes (**the most important part**) of the Squad OPORD and includes how the mission is conducted from the ORP to the OBJ. The best way to understand this is by understanding the procedures used to conduct certain missions such as Recon, Ambush, Movement to Contact, etc. It includes how you will move from the ORP to the OBJ, what order you occupy the objective, initiate the attack, and withdrawal from the OBJ. Since you are the decision maker on how to accomplish the mission, it is very important to pay close attention to what the Platoon OPORD stated. For example, it might state that the ambush must be initiated with an AT-4 regardless of size or density of an enemy unit. As a leader, if you don't think this makes sense, ask for a change to that part of the order. The platoon OPORD will also state the Priority Intelligence Requirements (PIR) for you to collect from the OBJ. Employment, by far is the most difficult paragraph to write but if you study the types of missions during the MS III year, you will be able to develop a quick concept and brief it with no problem. Additionally, when dividing the tasks for the mission, task each team with the requirements (e.g. A-Team you are the Assault Element, B-Team you are the Support & Security Element). At this point the element is task organized which assists the team leaders into taking initiative.

**Leadership Insight 28:** Keep the plan simple; don't develop an elaborate plan that is hard to understand. A simple plan executed violently is more successful than a complex plan that is never executed. Remember they acronym KISS (Keep It Simple Stupid).

***Phase 4 – Redeployment or Return movement***

This section is similar to Phase II (movement) but the only difference is it tells how the element will return from the ORP to the next AA or PB so follow the same format as Phase 2.

**EXAMPLE:**

*Once all equipment is accounted for at the ORP, we will be traveling by foot on a 120° azimuth for 200 meters to a checkpoint. From there we will travel on a 90° azimuth for 300 meters to the next AA. B-Team will be 1st in OOM using the traveling over watch movement technique. B-Team will be responsible for establishing enroute rally points every 200 meters.*

***Phase 5 – Post-Deployment or Post Mission***

Although the last phase of the operation is a small portion of the mission, it encompasses the AAR which can be a major learning opportunity allowing for improvement on future missions and maybe a better evaluation. Other activities include equipment maintenance, administrative actions and intelligence debriefings that are not addressed in the LDAC order unless the TAC adds information. Now that you have completed Phase III with all pertinent information and addressed every detail of the operation, it is time to move to the next paragraph.

The key to translating pertinent information from the *Service and Support* paragraph is to keep it simple and follow the Classes of Supply, identify the EPW and CCP for the mission and any other support required. The following Classes of Supply are usually addressed during SSTX: Class I (food and water) – tell everyone how many MREs they are to have and that they should have full canteens.

Class IV – Talk about how many rounds each soldier will have (e.g. Basic Load or the number of magazines), who is carrying the AT-4, who has demolition charges (Breach Team), what individuals have C-4 or claymore mines. These are the three main classes of supply you should address to get the basic information across. Let everyone know where you are going to put the Squad's CCP and EPW collection points. The best method is to make the EPW and CCP at the Rally Point (RP) or the Objective Rally Point (ORP). These two locations are known points so that the injured and EPWs can be consolidated until you are ready to move off the objective. Once you are commissioned, this paragraph becomes much more complicated and is the heart and soul of any military operation. LDAC does not address the importance of this paragraph very well, so understand what needs to be addressed and learn more once you are in the military.

The *Command and Signal* paragraph is very simple and is constructed by informing your squad of the higher commander's position (Platoon Leader), location of key personnel (Squad Leader), succession of command and SOP adjustments. When addressing the location of key personnel, address the squad leader's location during each phase of the operation. For example, during planning, the squad leader will be at the squad patrol base. During phase II and IV, the squad leader will be located between Alpha Team and Bravo Team. During phase III, the squad leaders location will depend on the mission. For example, if it is an ambush, the squad leader would be with the assault element, if it is a recon, the squad leader may be with R&S Team 1 and if it is a link up with a local national, the squad leader may be with the reception team. There is no right or wrong answer to the key leader position. However, use common sense on this part of the plan. Just as any great quarterback wants to have the ball when it is 4$^{th}$ and long with only 10 seconds left in the game, every leader wants to be where the action is to make the right decision. Therefore, it is your plan; place yourself where you can best maintain command and control of the squad. Most of this information is determined by you and very little gained from the platoon OPORD. When addressing the signal subparagraph, this information is almost exactly the same from the platoon OPORD, so pay close attention. Everything that you received from the platoon OPORD is pertinent for your squad OPORD.

Once the Platoon OPORD has been issued to the Cadet in charge of that SSTX lane, the TAC will ask if you have any questions. Use this time to clarify any information that was not clear to you during the OPORD or ask him to repeat information you missed. Next ask for any additional equipment that you might need to enhance your mission. Item such as binoculars, aerial imagery, machine guns, grenades, smoke or AT-4s.This will show that you are conceptualizing and understand what is needed to accomplish the mission. At this time you can ask for an adjacent unit to pull security for you during this point. You will need to be aware of what squads or units are to your left, right and what their missions are. For example, if 4th squad is reported to be in reserve, ask for 4th squad to pull security for your squad so you can give the OPORD without need for security from your own forces. Remember, notional security does not exist; you must use another squad or unit. Finally, sync your watch with the TAC's watch since all hard times will be according to his watch, not yours.

### The Squad WARNO and OPORD

Once the TAC issues the higher OPORD, return to the Squad Assembly Area to brief the OPORD and if there is time remaining from the original 10 minutes complete an effective WARNO by keeping it brief by issuing out basic information to include the situation, mission statement, general tasks and specific tasks. Many of these tasks may have been assigned during the Take Charge Minute but be sure to repeat it so the TAC understands that you can effectively disseminate a WARNO. At the end of the WARNO be sure to have your team leaders and RTO synchronize their watches to yours, which should be synchronized to the TAC's watch. This way everyone is on the same time line and your RTO can keep a running time hack for you throughout the mission. The RTO is your best friend and can keep you on time by providing you time hacks. Finally, WHAT EVER YOU DO, DON'T MISS YOUR LD TIME.

When giving the squad OPORD be sure to use the **5-PARAGRAPH OPORD FORMAT** at all cost!!!! This is very important as stated before and allows the information to be presented in an organized manner with a smooth flow without missing pertinent information. Before attending LDAC, you must practice giving the OPORD to members of your squad or your peers of the MS III Class. Briefing in front of a mirror, and using a terrain model will help you follow a logical sequence without looking at your notes when briefing the Execution Paragraph. When briefing, use a stick as a pointer to demonstrate surety and confidence and assist in briefing the plan to squad members. The pointer allows you to have an item in your hand for clarification and removes some of the jitters. Ensure that you tell the squad members to hold all questions until you are finished briefing. This controls the leader's peers and does not allow them to sharp shoot or derail the briefing process resulting in the evaluated Cadet displaying exceptional Military Bearing (MB).

**Field Graphics 101 - Terrain Model /Sand-Table**

After giving a quick WARNO, there are certain tasks that must happen simultaneously to meet the SSTX timeline. If task organized correctly, the RTO should be working on the terrain model, the Alpha and Bravo Team Leaders are disseminating information and conducting team rehearsals, while the leader is organizing the notes and thoughts to brief the plan. In a best-case scenario, there should be two terrain models oriented to the direction of movement and terrain. The first terrain model is for the squad leader to brief movement and the overall concept of the mission and the second terrain model is to brief Actions on the Objective (AOO) that is a magnified view of the objective area. The first sand table will have the route laid out as well as the AA, LD, primary & alternate routes, linear danger areas, rally points, ORP, and OBJ of the SSTX lane. The second terrain model goes much more in detail and is used to brief AOO. This terrain model should show the ORP, movement to the OBJ, security, assault and support positions, the OBJ, CCP, EPW CP and LOA. As you talk through your plan you will physically move the different elements across the objective to demonstrate the mechanics of the plan that takes place during the mission. Since we are here to assist you in your preparation for LDAC, we produced an awesome 43 piece Terrain Model Kit (TMK) that is waterproof, durable and long lasting to assist you in demonstrating professionalism.

Visit the website www.myleadersource.com to obtain the kit.

Once the OPORD is briefed, make sure to conduct a back brief by asking squad members certain questions about the information that was just briefed. For example ask a squad member what is the mission of the squad, ask the compass person

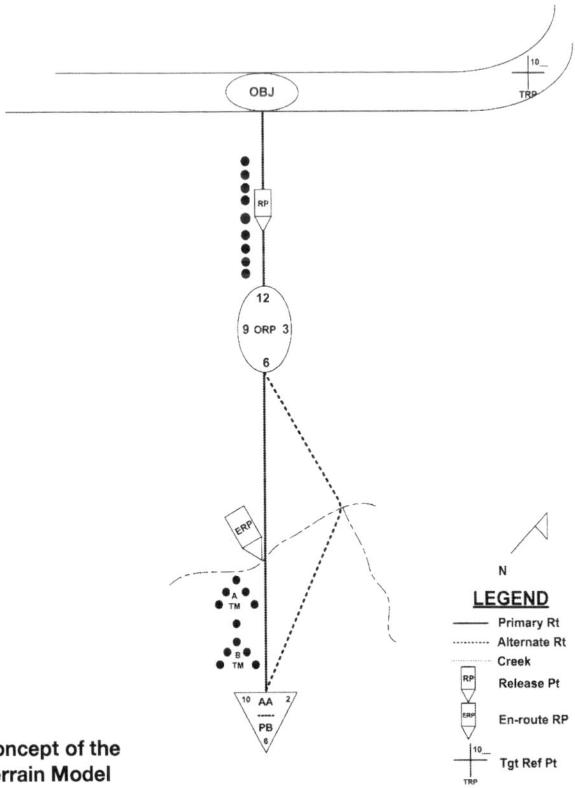


**Figure 18: Concept of the Operation Terrain Model**

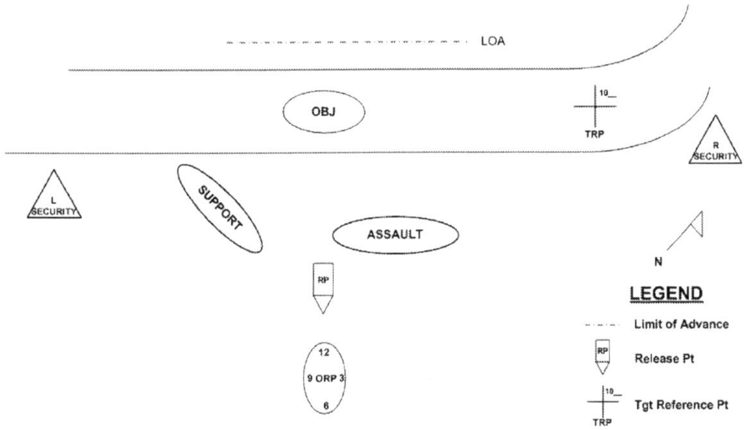


**Figure 19: Actions on Objective Terrain Model**

what is the azimuth of the primary route, ask the pace man what is the distance to the ORP and any other questions that you think are important such as the signal to initiate an ambush. Once back briefs are complete, ask the squad if they have any further questions. If there are questions, answer them in accordance with what you briefed and then move out to conduct rehearsals.

**CONDUCT Rehearsals** at all costs! The last 15 minutes of the planning phase is dedicated for rehearsals which can make a mission very successful after a weak OPORD or place the finishing touches to a great OPORD. The purpose of rehearsals is to allow the squad to understand the mission fully by walking through the required actions to assist in visualizing what to expect. At a minimum, rehearse the leader's recon, how elements occupy the objective, actions on the objective and the withdrawal from the objective. If you have more time, rehearse movement to the objective, battle drills and contingencies. During rehearsals you will explain again what you want to happen while walking through the process with the squad doing an abbreviated version of the mission. This reinforces the step by step brief of what should happen, including identifying key Ankle Biters before they surface.

The Line of Departure (LD) is approximately the half way point of your evaluation and is another crucial time because the higher OPORD specifies that units must cross the LD NLT H+45 for SSTX and H + 90 for PSTX. Additionally, the element must contact higher headquarters before leaving the AA. Most Cadets fail to document the time they received the OPORD and some have a very difficult time converting the H+45 to an actual time base off the time they received the platoon OPORD which causes them to miss their LD time. If you miss LD time or cross the LD early, you will more than likely receive an S unless the remainder of the mission goes flawlessly. Once you are ready to depart, request permission to depart the AA and wait for confirmation from the TAC. Failure to conduct rehearsals, failure to contact the TAC prior to crossing the LD and departing the AA early or late earns the evaluated Cadet multiple demerits. Once you have finished rehearsals, do one last PCI for your squad using your team leaders and allow the compass man to get an accurate azimuth. Next radio higher using proper procedures the proper call sign from the platoon OPORD and request permission to cross the LD. Finally, no matter what,

**DON'T MISS THE LD TIME!**

**EXAMPLE:**

*CADET says: Tango 1-6 this is Charlie 3-1 request permission to cross the Lima-Delta. Over.*

*HIGHER HQ says: Charlie 3-1, permission is granted to cross the Lima-Delta. Tango 1-6 Out.*

**Leadership Insight 29:** Don't cross the LD early or late; use rehearsals from the time you complete the OPORD until it is time to cross the LD.

**Movement** spans across two different phases when conducting field operations and begins when the squad or platoon cross the LD and ends when the element arrives at the Objective Rally Point (ORP). There are many things to consider during movement since this is a vulnerable point of the operation. You must use the proper squad formation and movement technique that provides the best security depending on METT-TC and the most firepower or speed to accomplish the mission depending on the situation. Depending on METT-TC, the leader may opt to use either the squad column-fire team wedge or the squad file. The first formation provides the most security during movement and the second provides the most speed during movement. A squad line places security and the maximum amount of firepower to the front but tends to be the slowest. Once the best formation is determined for the environment and terrain, the next decision is how to place the squad members within that formation which is very flexible. No one is going to direct a Cadet to use a specific formation at LDAC but it will be very apparent to the Evaluator if you understand how to employ movement techniques tactically. Bottom line is to make a decision and if it does not make sense issue a FRAGO; remember the Golden Diamond Process! When moving, ensure that the point person is establishing enroute rally points approximately every 200 meters just in case the squad is attacked. The enroute rally point becomes very important for consolidation and reorganization if attacked. The two most used formations during field phase are:

The FILE Formation - Squad members' position within the file should have their weapon oriented in an alternating direction for maximum security and adjust the distance between personnel depending on the terrain. The Squad Automatic Weapon (SAW) is assigned to each fire team and should be placed on different sides of the formation while moving. Although Alpha Team is primarily the first in the order of movement, Bravo Team can be placed in the lead with no issue. The Squad Leader and RTO are in the middle of the formation for the best command and control. Each person is responsible for a certain area that overlaps with the soldier next to him in order to provide 360 security. The rear most individual has the largest field of vision and must provide rear security during movement.

The LINE formation – Is used when you want the most firepower in the direction of movement or when assaulting the objective from the last cover and concealed area. Although massive firepower is given to the front, the squad is very vulnerable to the rear. For this reason, squad members are given secondary fields of vision in which they turn around to view the rear every few steps or meters. The key is to keep your "head on a swivel" which means constantly scanning the area for threats.

**Leadership Insight 30:** The squad column fire team wedge or the file formation is used the majority of the time during SSTX lanes.

The purpose of an Objective Rally Point (ORP) is to establish a location that has cover and concealment, can be defended for a short period of time, has a location that is easy to find and has close proximity to the objective for ease of control. The ORP is also used as a staging point to conduct a leader's recon, stage rucksacks, prepare equipment that will be used on the objective and as a point to link up after kicking some ass on the objective. There are two methods to establishing an ORP: 1) conduct a leader's recon of the area and then establish the ORP or 2) take the ORP by force without a leader's recon. Due to the time constraints of a 2-hour SSTX at LDAC, Cadets are advised to take the ORP by force. This involves moving along your primary or alternate route until you are in a location to establish the ORP which is determined in the planning phase. Once you arrive at the ORP, complete the following tasks:

Establish a cigar shape perimeter with 360 degree security – Recommend placing personnel approximately 5 -10 meters apart and give them sectors of fire. Ensure there are no gaps in between positions. Be flexible on the cigar shape perimeter because it is possible SSTX and PSTX may not be on the same sheet of music. This is a great time to "Manage your Boss" and see what the TAC's pet peeves are on this subject; do they like the cigar shape perimeter or do they prefer something different? Figure it out and adjust your decision.

Issue a **GOTWA**! This is nothing more than a simple no communication plan that encompasses 5-points of a contingency plan to synchronize two separate elements for a future link-up. GOTWA is an acronym that has five pieces of information that represents necessary information.

G – Where you are going

O – Others you're taking with you

T – Time you will return

W – What to do if you don't return or time elapses

A – Actions of personnel if they come in contact with the enemy; (2-parts); what personnel in the ORP do and what the leader recon element does if attacked.

*For Example:*

*G- We're going on the leader's recon.*

*O- We're taking the RTO, the Alpha Team Leader and the Alpha team rifleman.*

*T- We'll return in approximately 30 min.*

*W- If we don't return in 1 hour, abort the mission, fall back to the last known Rally point and call higher to inform them that there is a break in contact and ask for further instructions.*

*A- If the ORP is attacked; if element is larger than ORP personnel, break contact & move back to the last known rally point, if it is smaller, eliminate the threat and be prepared to move to the last known rally point; if the Leader's Recon comes into contact, we will break contact and move back to the ORP.*

A **Leader's Recon** is conducted once you occupy the ORP. At that time, Team Leaders should automatically send personnel going on the leader's recon to the center of the ORP once security is established. At a minimum, the RTO, Alpha Team Leader and a compass person should accompany the squad leader on the leader's recon. (Of course this is dependent on the leader's plan) The leader issues a 5-point contingency plan (GOTWA) to everyone going on the leader's recon to include the Bravo Team Leader and immediately following the GOTWA, the Bravo Team Leader counts personnel out of the ORP at the 12 o'clock position as the element exits to conduct the recon. The Bravo Team Leader then briefs the remaining personnel on the GOTWA, adjusts the perimeter so that 360 ° security is maintained and conducts necessary tasks to prepare for the mission. The personnel that make up the leader's recon for a squad size mission should be approximately three to four personnel. By taking four personnel, the squad leader has the flexibility to leave surveillance or security at the objective while the squad leader and security member return to the ORP to lead everyone else back to the objective. After everyone is informed, a rucksack plan is established so personnel can retrieve their rucksacks very quickly after the mission. The rucksack plan development is up to the squad leader but the best method is to place the rucksacks in two rows with the shoulder straps laid on the ground so that each member can retrieve any rucksack upon returning to the ORP.

During the Leader's Recon, there are several pieces of information that are mission dependent that a leader must obtain. The most common pieces of information that you are looking for are: 1) Release Point 2) Identify the objective 3) Key element positions such as assault, security & support 4) Limit of Advance (LOA). For example, if you are going to Knock out a Bunker, a leader wants to find the ideal place for the heavy weapons position and identify the last covered and concealed positions before attacking the bunker, the best avenues to enter and exit and a controlled withdraw from the objective. One fatal mistake that Cadets make during the leader's recon is they don't assess the killzone or objective very well because they hide in the weeds with tunnel vision and fail to actually see the objective. Don't be a bone head; if security is established, then get up and assess the objective and finalize the plan. If you are attacked, implement the GOTWA and get evaluated on your actions; it is not a big deal if you execute violently (You are being evaluated on your actions as a leader).

Once the leader's recon is completed, it is time to return to the ORP and disseminate any new information or changes to the plan; another great chance to implement the Golden Diamond Process. Since time matters at each SSTX / PSTX lanes, the squad leader must be really quick to get the new information out to everyone! Therefore, it does not take a tactical discussion (a full blown planning process again) to brief the updates! Keep it simple and very basic.

**Actions on the Objective (AOO)** is where a leader can make or break an evaluation during SSTX or PSTX lanes. There are many different missions that you will be tasked with at LDAC and each one has its own set of procedures and actions on the objective but most of them follow these three easy guidelines: 1) identify the threat 2) eliminate the threat and 3) continue the mission. The different types of missions that will be conducted are Small Unit Tactics (SUT) derived from FM 7-8; examples are an ambush, recon, movement to contact, clear a minefield, knock out a bunker, squad attack or react to contact. The actual type of mission is not that important because Cadets must be knowledgeable of the procedures on each of these missions in order to follow the three guidelines stated earlier. The threat may be an enemy recon patrol, it could be detection from the enemy, maybe it is a minefield, or maybe it is building rapport with a local national. Either way, you must realize that these obstacles are a threat to your success and therefore you must identify and eliminate the threat and continue the mission before reaching that success. After the threat has been identified, there are multiple ways to eliminate it. It could be by neutralizing the enemy recon patrol, placing controls to avoid detection such as using binoculars and camouflage or it could be using a mine detector to clear a lane through the minefield. Eliminating the threat is removing the barriers that are between you and mission success. Finally, continuing the mission requires Cadets to remain focused! For instance, if a Cadet just neutralized an enemy recon patrol during the ambush, the next step in continuing the mission is to secure the objective and search the enemy. If you encounter an enemy patrol during the movement phase, the next step is to neutralize the enemy patrol and continue your movement to your original objective. As you can see, it is all about being flexible and taking charge! You must be vocal, confident, and conceptual to take full advantage of these opportunities.

*Example of procedures for Actions On the Objective (AOO) during an Ambush*

*Initiate the ambush with most casualty producing weapon (e.g. M240 or claymore)*

*Make the decision to assault across the objective*

*Secure the objective*

*Receive ACE Report (Ammunition, Casualty and Equipment status)*

*Control the flow of specialty teams onto and off the objective (e.g. EPW teams, Aid & Litter teams, Demo teams, CB teams). It is up to the Cadet in charge to assign specialty teams for the mission. An example of specialty teams responsibility is shown on the following page:*

| ***Alpha Team*** | ***Bravo Team*** |
|---|---|
| *Primary EPW Search* | *Primary Aid & Litter* |
| *Primary Pace & Compass* | *Primary Demo* |
| *Primary Recon & Security* | *Primary Bunker Clearing* |
| *Alternate Bunker Clearing* | *Alternate Pace & Compass* |
| *Alternate Demo* | *Alternate Recon & Security* |
| *Near Side Security* | *Far Side Security* |

*Implement a contingency plan if necessary*
*Eliminate problems as they are presented*
*Call in necessary reports (SITREP, MEDEVAC, ACE)*

A **Withdrawal Plan** is a must after completing necessary actions on the objective; the squad or platoon must return back to the ORP in an orderly fashion to maintain control and efficiency. If all goes to plan, then the squad leader will implement a planned withdrawal. If the mission goes to hell in a hand basket then chaos will prevail without a plan. Either way, a plan for both is necessary! If all goes to plan as it will the majority of the time, the squad leader will initiate the movement verbally. If you are conducting an ambush, there are three elements that must be withdrawn from the OBJ; Assault, Support and Security elements. All of these elements will withdraw through the Release Point after receiving a verbal order or signal such as a whistle where a team leader will count them during the withdrawal. For example, during the withdrawal of an Ambush mission the Assault element withdraws first, the Support element second and finally the Security element. Since one of the team leaders count personnel off the objective the first person back to the ORP will count personnel into the ORP until a team leader arrives to ensure accountability. Upon arrival to the ORP, squad personnel retrieve their rucksacks and prepare to depart the ORP as if the squad was returning to the AA or moving to a new patrol base in order to prepare for another mission. This is great hip pocket training to conduct in Garrison Operations or at the FOB. Most of the time, the TAC will stop the mission and inform you to get your equipment from the ORP and meet at an After Action Review (AAR) site. However, don't become complacent and blow your evaluation by not remaining tactical, motivated and maintaining control of your squad. If you waste too much time, it could affect your evaluation. Conduct an AAR and be ready to receive your evaluation.

Currently at LDAC, you or the Cadre conduct an **After Actions Review (AAR)** but this may change from year to year. If you are in charge, then you are the primary orchestrator who will facilitate the AAR by asking pertinent questions and controlling the tempo and topics. A good technique is to have the team leaders stand up in front of the squad and address the operation by the phases in the OPORD. Start by stating the mission of the squad and then ask the Alpha Team Leader what he or she thought about the planning phase of the operation. Let the Team Leader talk and then ask members of the Alpha Team if they have anything to add such as one positive and one negative remark. Get everyone involved so that the AAR becomes constructive. After three comments (guideline) are discussed, move to the next phase and ask the Bravo Team Leader what he or she thought about the movement phase and then continue the process as stated before. Once you have a couple of points of sustain and improve for that phase, move on to the next phase. Finally, keep it interesting and relative to the mission and don't allow it to become an argument. If a peer tries to take over your AAR, cut him or her off professionally and move on to the next point. Some Cadets will attempt to Monday morning quarterback the operation to show how smart they are and how inept you are. Be prepared to react to this behavior with a calm and political approach.

Finally, many great techniques and methods have been discussed in the preceding text on how to get an outstanding evaluation during a SSTX or PSTX lane. However, you are the LEADER who must apply those techniques and methods when in charge! A good leader observes other leaders when they are not in charge and learns from their mistakes without making the same mistakes. As you learned earlier, Cadets must learn to read their Evaluators to see what the Evaluator's pet peeves are to keep from making the same mistakes again. Some Cadets will find this very easy to accomplish and others will find it very difficult. With that said, every Cadet can achieve success if he or she prepares before arriving at LDAC or is a quick learner. If all else fails, emulate success; watch your peers. All ROTC programs do not put the same emphasis on the same type of leadership evaluations, which is the reason LDAC is designed to evaluate the total leader by placing Cadets in different situations to evaluate their leadership skills. For example, some programs focus on Garrison Operations and some focus on Field Operations; therefore a Cadet may be really strong in one type of evaluation and weak in another. The bottom line is when in charge, TAKE CHARGE! When not in charge, be a good follower and don't make the same mistakes as the Cadet before you made.

# 11 Basic Living at LDAC

As a leader and a follower, there are a few motivational items such as food and mail that affect your operations, effectiveness and morale. As a leader you are expected to take care of your subordinates, with the ability to coordinate and ensure that they have the material and services to carry out operations. Material and services are made up of 10 classes of supply, transportation, laundry and services, maintenance, medical evacuation and personnel support. You may not address each and every one of these items during each leadership evaluation but you will experience all of them at some point during your time in LDAC-I-STAN.

*Class I (Ration Plan).* At LDAC if you do not develop a plan to feed your soldiers, their morale and ability to conduct operations will be affected, which will affect how they work for you and conduct operations. During Garrison Operations Cadets usually eat breakfast and dinner in a Dining Facility that has predetermined times. Every Cadet in your Regiment (almost 400 Cadets) eats in this dining facility, which requires reverse planning to ensure your platoon is in and out in fifteen minutes. Although it is up to the Cadets to develop organized procedures and discipline for processing their members through the dining facility, there are certain rituals that must take place, such as everyone performing pull ups before entering. For lunch you will be issued Meals Ready to Eat (MRE) that will be consumed at the training site for the day. Rest assured that Cadets are being evaluated during this process.

During Field Operations, Cadets eat MREs during the entire duration so if you are in a leadership position, you are responsible for requesting these meals for each day of training. If an MRE request is not submitted through the Cadre and supply system, the MREs may not arrive, causing the entire platoon to be very hungry and not work for you during the leadership. Since LDAC is attempting to replicate the same procedures as in a regular unit, the Cadre may not bring this to the leader's attention until the last moment, which will cause a situation that in turn will create a behavior that can be evaluated. If I were evaluating you and you failed to request Class I, I would give you a negative mark on my notes and it would be documented under the innovative (IN) and Domain Knowledge (DK) on a Blue Card, however, it's not the end of the world. Then again, I may use this instance to see how you recover from the situation. This may end up in getting positive marks in other dimensions because you adjusted to the situation and took charge. Another key note is that there is no dedicated time to eat MREs in the field. Therefore, Cadets should field strip their MREs and eat when practical to assist the current leadership.

*Class II (Clothing).* The main time that you will use this is when you draw or turn in items from the Central Issuing Facility (CIF) at LDAC. There are other issues that fall under Class II but you will not experience them at LDAC.

*Class III (Petroleum, oils and lubricants).* You will not be using this class of supply at LDAC; however if you find yourself in a Tank Unit, this is very important because you may become a sitting duck if you fail to plan for a resupply.

*Class IV (Fortification and Barrier Materials).* You will not be using this class at LDAC either due to resource constraints but an occupation of a FOB will require extensive materials for protection against the enemy. However, you should still request the resources if necessary to show that you are tuned in.

*Class V (Ammunition).* This will become one of the most important requests that you make while at SSTX and PSTX. Each time leadership changes, that leader will be responsible for conducting an inventory during the changeover. Conduct the inventory and get an ammunition status of your squad or platoon to ensure that each of your soldiers has the appropriate ammunition to conduct the next day's missions. Always be ready to cross-level to ensure everyone has ammunition.

*Class VI (Personal Items).* Think Shopette and phone use! Every good leader in the Regular Army has a plan for resupply operations for the ten classes of supply and at LDAC, Cadets are in charge so it is their responsibility to plan for resupply operations. Some Cadets arrive at LDAC without the proper personal items needed for their month long adventure and of course there are some Cadets that must replace personal hygiene items because they only brought a half a tube of toothpaste. Another issue is that females have specific needs to ensure they can perform proper field hygiene; however, this does not mean they get special treatment. Females know what must be done on their behalf and if they don't, have another female in the platoon educate them before departing Garrison Operations. Therefore, when you are in charge, treat the Shopette runs as a resupply operation to refit with necessary items. Since we are talking very minimum items, it does not mean that everyone in the platoon needs to go and take the Shopette Tour to get out of work. Have each squad designate two individuals to conduct the resupply for the team so that the platoon operations do not suffer. Another sensitive area that is not necessarily Class VI is the phones; however, I would treat this as a personal item because Cadets want to use the phone to talk to their loved ones. If you don't have a plan for phone use, you will lose control and they will all migrate to the phones anyway. You must plan for these two personal requirements (training distractions) or your peers will take advantage of you during your leadership position causing tasks to be sacrificed as well as your leadership evaluation!

*Class VII (Major End Items [e.g. weapons]).* The only major end item that you will deal with at LDAC is your individual weapons. You will have certain days that weapons must be drawn from the Arms Room so take a look at the training ahead of time to see if there is time allocated for the task. If so, the XO must coordinate through your Company Cadre and Cadet Chain of Command and should consider this an Ankle Biter at LDAC because leaders are being evaluated on how the weapons draw and turn-in is conducted.

*Class VIII (Medical Supplies).* Most Cadets confuse this with the Medical Evacuation Plan discussed later in this section. This request is generated when the M5 aid bag must be resupplied or just prior to attending the First Aid Training. The M5 bag assist in conducting medical classes as hip pocket training between training events. There are specific days on the training schedule that will focus leaders on when to conduct this medical training, so be proactive and don't wait for someone to tell you to conduct the training it is on the training schedule and briefed in the OPORD process as training focus of the day.

*Class IX (Repair Parts).* In the Regular Army, this is a major operation to get repair parts to the unit. At LDAC, the repair parts will be for your equipment that has to be replaced such as a rucksack frame. Direct Exchange (DX) of the equipment is necessary so that the unit can accomplish the mission so take charge and assign a supply representative to schedule the DX of equipment through the Platoon TAC Officer.

**Transportation** at LDAC is the Center of Gravity and is treated as such. Remember the term "Cadre Override"? Fail to make transportation times and the Cadre Override Button is pushed with immediate action! Transportation for each day of training is coordinated before Cadets arrive at LDAC and is a very systematic process that has no wiggle room for delays. Your arrival to a specific training area is crucial and relies on the Cadet leadership to ensure personnel with required equipment are loaded on each vehicle in a timely manner. Remember that a delay of only a few minutes creates second and third order affects that influence training, your leadership evaluation and other Regiments' training schedules. Although the transportation has been coordinated, Evaluators use this daily requirement as a coordination event to evaluate behavior, therefore if you are the PSG / PL, transportation should be coordinated through the Company Cadet XO and then he or she will coordinate the transportation for the Regiment through the Cadre. If you are the Platoon Leader, you are responsible for developing a transportation plan for your platoon that may incorporate two types of vehicles; either buses or silver-side trailers. Cadet leadership must determine how many Cadets can be placed on each vehicle and how the leadership can maintain control during movement. My recommendation is to develop a simple Transportation Annex to your OPORD and disseminate it during the brief to be effective.

**COLD HARD FACT 17:** If you miss movement, you will experience an emotional event called "Cadre Override" and it will not be fun! The result will be a negative mark on your leadership evaluation.

**Maintenance and Services** at LDAC is a Cadet's individual responsibility since they are away from home and not living with their mothers. There are washing machines, dryers and male showers located in each of the barracks; female showers are located in a separate building outside of the living barracks. These activities will be accomplished after training and before you go to bed at 2200 hours so developing a schedule is a necessity so that everyone has the time to wash his or her clothes and shower each day. Believe it or not, there are Cadets who will not take showers on a daily basis because of two reasons: 1) leaders do not disseminate information which results in wasted time and disorganization or 2) the Cadets want to be like their favorite Peanuts cartoon character "Pig Pen." Either way this becomes a personal hygiene issue and can affect the health of the entire platoon and can single out the Cadet. Some type of maintenance is conducted every day at LDAC. It is a must to maintain individual, squad and platoon equipment so it will function properly when you need it. This equipment includes items such as your Load Bearing Equipment (LBE), Kevlar, Rucksack and Weapon. Items like your rucksack straps must be maintained to work properly during foot marches, your rucksack will affect your performance and become a nuisance. Weapons maintenance must be performed regularly to ensure it operates properly when needed. This is an individual task but leaders must ensure that it is carried out. During fire guard, personnel can transfer laundry from the washing machine to the dryer to assist a Cadet in time management and be prepared for the next day's training. One key note is that Cadets are not allowed to be up past lights out unless they are using the latrine or are on fire guard; by the way, the washer and dryer are located in the latrine.

During your MS III year you will work with the same Cadets on a regular basis and will formulate **Standard Operating Procedures (SOPs)** at your home campus that will make your unit more efficient during Labs and Field Training Exercises (FTX). Most ROTC Brigades conduct a combined Brigade FTX prior to MS III Cadets attending LDAC in the spring semester, which is usually the first time Cadets are exposed to Cadets outside of their schools, requiring SOPs to be adjusted because every school has different SOPs. LDAC is very similar and will require SOPs to be adjusted during Garrison Operations. Each squad will have some time during Garrison Operations to develop and adjust SOPs so that the element is more efficient at SSTX and PSTXs. Most squads fail to develop these procedures prior to moving into field operations because they blow them off or are overwhelmed by a series of events. The bottom line is that each day your platoon will have TAC time or some other free time during training events to develop these SOPs to prevent chaos. As platoon leadership, it is your responsibility to ensure that squads are conducting opportunity or hip pocket training during this down time. As a Squad Leader this becomes a great opportunity to develop procedures that will assist everyone's leadership evaluations in the field and reduce the amount of chaos. This is very similar to a regular unit because Squad Leaders utilize this same opportunity training to ensure their squads are trained and prepared to conduct combat operations. Platoon Leaders and Platoon Sergeants oversee this training and use it to ensure that collective training supports their Mission Essential Task List (METL). At

LDAC your evaluations are based on the same day-to-day and field tasks required to accomplish operations, therefore, ensuring this training takes place will produce great marks on the Blue Card Evaluation. If you are unsure of what type of training to conduct, reference the daily training schedule for training focus of the day. If you are the Platoon Leader and Platoon Sergeant, you should address the Training Focus in your platoon OPORD under intent, guidance or coordinating instructions. Below are some opportunity training subjects for SOP development:

Movement Techniques

Hand and Arm Signals

First Aid Procedures to include Casualty Evacuation

Pre Marksmanship Instruction (PMI)

Rucksack Recovery Plan

Enemy POW Searches

Squad Terrain Model

# 12 Medical Issues, Blisters and Boots

The most significant medical issue at LDAC is foot injuries and the best way to mitigate these injuries is through prevention and foot care education prior to attending LDAC. Many Cadets arrive at LDAC without the proper physical preparation and stamina required to negotiate the rigor of continuous training for the length of the course. Couple this with an injury and you have all the ingredients for negative impact towards training and the Cadet's performance. Failure to have a properly fitted pair of boots creates BLISTERS which become the most common foot injury at LDAC-I-STAN. When blisters form, Cadets' performance suffers because they become overwhelmed by THS; being trained in proper foot maintenance and prevention becomes a necessity to avoid becoming a liability.

**COLD HARD FACT 18:** If you get blisters or an injury at LDAC, the Cadre and fellow Cadets are **NOT** going to feel sorry for you! They expect that you pull your share of the weight and drive on without whining or trying to get over by using the injury as an excuse. Evaluators are **NOT** going to treat you any differently or give you a break because you are injured. If your injury is not severe enough for a medical recycle or a return trip to home campus then you are fit for training and will be evaluated as if you were at 100%.

Don't try to BS the Evaluators or your peers because it is fairly cut-n-dry if an MS III Cadet has prepared for LDAC or not. In order to be prepared for the rigor of LDAC, Cadets must have two pairs of well-fitted and broken-in boots by the start of spring semester. They must be physically fit and be capable of performing foot marches up to five miles with a thirty-five pound rucksack bimonthly before arriving at Ft. Lewis. The proper fitting, lacing and maintenance of the boots, blister and abrasion treatment and knowing how to maintain your feet properly are all very important issues that must be addressed and understood to allow Cadets to make it out of LDAC-I-STAN with all of their toes.

*Proper fitting* of a pair of boots takes two important areas into consideration: length and width. When fitting a pair of boots, the space between the ends of the big toe of the boot should be the width of the thumb and there should be sufficient space under the lower edge of the tongue to admit an index finger in an unlaced boot. Improperly fitted boots can cause blisters, abrasions, calluses and corns; usually from boots being too small thus causing friction or pressure. Over the past twenty years of service, I learned how to take care of my feet to ensure they never failed me but even then there are times when boots break down and do not protect the feet as they should. When Cadets join ROTC they are given two pairs of boots that they will wear to conduct military training, including LDAC. Since the military has hundreds of thousands of people, they buy boots that are common sizes that are designed to fit a large population. As we all know, this may not be the best solution for certain individuals if they have a unique foot shape or structure, which can become a major problem for the individual if the common sizes do not fit correctly. Although the military does not have a custom boot fitting program, Soldiers can go to a military podiatrist for counsel and if deemed necessary, a prescription can be given to obtain custom fitted boots. For the most part, the majority of people will be able to wear the common sizes. There are a thousand different techniques to fit boots properly but after researching several manuals and websites, the best description I found was from The Great Outdoors website located at:

***http://www.greatoutdoors.com/published/boot-fitting-guide***
They have listed several Rules of Thumb to assist troubleshooting and determining the causes of friction or pressure. Although I am not aware if The Great Outdoors specializes in manufacturing custom boots for the military, they are an excellent place to purchase additional equipment for deployments and your personal recreation hobbies and a great source of information. Visit their website for more information, detailed instructions and photos of lacing.

**Boot Conditioning and Preparation** plays a major role in ensuring the feet are ready for any mission or long walks. Feet under the stress of a combat load will become wider, longer and will swell approximately 1% - 2% during forced marches. Best-case scenario is to go out and run two or three miles with the insoles and socks you will use to determine if they actually fit. Insoles provide relief to the shock generated by heavy loads and should be worn during the early stages of a road march. If desired, the insoles can be removed or replaced with thinner insoles to allow for more comfort when feet swell. Normally, Cadets sign for their boots from supply and never fit their boots until it is time to wear them, which can result in hurting their feet. My best advice is to conduct the above test after you receive the boots and if they do not fit, return them for another pair. Whatever you do, don't keep the boots if they do not fit. If your supply tech tells you to deal with it, talk to your primary instructor for a second opinion.

**Leadership Insight 31:** Generally, if you have a good boot fit, you should have a slight amount of room in the front of the boots to wiggle your toes. Width-wise, your boots should feel a little snug (but not tight) and your heel should stay in place while walking and not slide back and forth (which causes blisters). To improve the fit of the boot, try loosening or tightening the laces at different parts of the boot until you get the perfect fit.

*Boot Lacing and Boot Care* - Proper lacing of boots not only prevents blisters but it will prevent improper circulation in the foot. Since some laces may assume a seesaw action across the instep and produce a long blister across the instep, I recommend skipping the lace over the instep to prevent irritation, resulting in a more comfortable fit. If you are like me, it is very hard to get a boot that fits properly without manipulating the design to accommodate your foot. Always carry an extra pair of laces in your rucksack because they will break at the most inopportune time. Remember, an improperly laced boot can cause injury and blisters that can affect your performance.

The Army Desert Tan Boots are made of suede and cordura nylon, which are two very different portions of materials and both must be maintained in order to perform to standards and not fail you. When caring for the nylon fabric, a soft brush with warm water can be used, whereas the suede portion only requires a soft bristle brush to remove the dirt; however, neither area of this boot should ever have polish applied. The boots should be air dried and **NEVER** exposed to direct intense heat (e.g. fire) because this can cause damage to the boots.

**Boot Foot Guide**

*Rules of Thumb:*

- If the tops of the toes are involved, the cap is too low or too stiff.
- If the ends of the toes are affected, the boot is too short or too loosely laced.
- If the sides of the big and little toes become irritated, the

boot is too narrow.

- If there is irritation at the heel, the boot may be too long, too loosely laced or have too wide a heel space for the foot.

*Heel* - If your heel is loose, try tightening the laces in the area directly at the ankle-bend.

*Instep* - If there is too much pressure over your instep (the top of your foot near your ankle), try completely skipping one set of eyelets with the laces, directly over the affected area. One drawback to be aware of is that you may experience a looser heel with shoes laced this way.

*Wide Forefoot (ball)* - People with wide feet (in a boot that is too narrow) often feel slight pain or numbness at the ball of the foot. To fix this, first try loosening the laces (in the affected area only). If this doesn't work you may need to try a thinner insole, which will give your feet more room. Also, check out the boot-freezing trick for Extra W-I-D-E feet. Narrow Forefoot (ball) - People with narrow feet (in a boot that is too wide) can try two things: Tightening up the laces often works, but once the boots have stretched slightly you may run out of room for more lace tightening, which then will require a thicker after-market insole to take out the excess space. Try a flat (arch-less) insole for this purpose, and put it underneath the removable arch insole that came with the boot. If your narrow foot is causing heel slippage, use the Runner's Knot.

*Pressure on top of toes* - Boots that crease abnormally and dig into the tops of the toes are too wide, too long, or have too much interior volume for your foot. Thicker insoles may help!

*Loose Heel, Narrow Foot* - The Runner's Knot can eliminate heel slippage in most cases but be sure to snug the slack out of all the loops in the hitch before giving it your preferred tightening tension. (Important! Pull down toward the ground to remove slack. Then pull up, and cross the laces to get your preferred tightening tension. Finish lacing the eyelets in normal fashion.) This knot may present problems for hikers with high insteps by putting too much pressure on the top of the foot. See the Instep Hitch below if you have a high instep.

*All-around Custom Lacing* - The Surgeon's Knot is great to use with all sorts of lacing techniques because it does not slip easily. To tie the knot, cross the lace three times, then make a loop out of each remaining lace to form what looks like two bunny ears. Keep one of the loops stationary like a tree and then have the other loop go around the tree and then push that loop through the hole that was just made and pull tight. This knot is best used as a "lock," such as when you need one section of your laces tighter (or looser) than another; if you have wide feet the lower laces can be left loose for more room. The surgeon's knot can be applied to the eyelets near the bend of your ankle and the remaining laces in the upper eyelets can be secured normally.

*Extra W-I-D-E Feet* - Freeze them! Not your feet but the boots, of course! Take two puncture-free produce bags commonly found in a grocery store. Place one bag inside the other.

Stuff these doubled-up bags down into your boot, as if you were trying to line the boot with the bags. Lace the boots up with a lacing tension similar to what you would use if your feet were in them. Now carefully add water to the produce bag and avoid getting water in the boot lining. When the bag contains enough water to fill the entire boot area below the ankle, securely tie off the top of the bag using several wire twist-ties. Now place the boot, bags, and water into the freezer, where the liquid will solidify overnight and expand in order to widen the forefoot of your boot so that a D becomes a E width. If you need to drastically widen the boots, use this procedure in multiple stages, using new bags for each width stage.

*SOCK Guidance* can be a very long discussion but I will sum it up in a short passage since there are numerous systems on the market. We will talk about three different systems: Army issued wool socks, Synthetics and Smart Wool socks. The Army issued wool sock is a bit dated but still works if it is the only system available. However, if the Army issued wool sock is used, ensure a two-sock system is implemented to support heavy marching. The inner sock should be a thin "wicking" material made of cool-max or polypropylene to remove moisture from the foot and transport it to the boot. Another option is to wear women's knee high panty hose as the inner sock. Although you may laugh and think this is a bit feminine, it works! The second system and much better than the Army wool socks are synthetics. They last longer, provide superior fit and are much better at dealing with moisture than wool. Synthetic socks wick moisture away from your foot after it condenses into a liquid form, resulting in drier feet. Although abrasion and friction is not as noticeable with synthetic socks, spend the extra money to buy the additional liners for the following reasons: 1) Liner socks can cure some "problem fits," especially those caused by narrow feet 2) Liner socks are part of a two-layer system, which is an excellent way to deal with lots of moisture 3) Liner socks can also be used on long trips when you want to take along a minimum number of socks. The major difference between the Army wool socks and the synthetic sock is synthetic socks dries quicker, last longer and act as a liner against your feet, saving the outer socks from the damage of sweat and oils. The last system discussed is Smart Wool made from merino wool fibers that assist in keeping feet warm in the winter and cool in the summer. This is accomplished by forming tiny air pockets that act as a barrier between the cold and heat. The Smart Wool socks differ from the synthetic socks because they transfer the moisture away from the foot in the vapor stage before it becomes a liquid. Even though I have worn all three systems, and all three system work, in my opinion, Smart Wool socks are the best on the market. They are a bit pricy but how much are your feet worth? At LDAC, there will be times when wet conditions exist every day and it may seem as if your feet will never be dry. This becomes a major issue if Cadets do not change their socks and use foot powder because trench foot can surface very quickly that may result in not completing LDAC. As stated before, all three systems work fine but each and every individual have different preferences and limitations so try all three and figure out which one works best for you before investing a lot of money into socks. However, no matter what the cost, take care of your feet because you will be on them for many hours each day.

**Leadership Insight 32:** Change your socks at LDAC on a regular basis and use foot powder. Your feet will get wet at LDAC so changing your socks is a must and allows you to keep the second pair of boots dry once you arrive at the Assembly Area (AA) or Forward Operational Base (FOB).

*Foot Preparation* - The best way to prevent blisters or a foot injury at LDAC is to have proper fitting boots, toughen up your feet and practice proper maintenance before arriving. Start the preparation process at the beginning of your MS III year in order to have plenty of time to accomplish the great feat of foot preparation; it is a necessary evil! Some Cadets get pedicures but unless you specify to the foot specialist that you are toughening or conditioning your feet, they will remove the toughness so that your feet are baby skin soft. Therefore, avoid pedicures or educate your foot specialist! Pedicures can be one of the worst things you can do before attending LDAC.

After several weeks of training, your feet naturally toughen up and the locations where you originally experienced blisters will no longer get them and the skin will get rough. While the general effect is good, there is a component of this that is not so good—calluses. Callused skin is too rough and should be removed because calluses in the heat can separate from your foot and blister underneath or even come off completely, leaving the foot raw underneath.

*I know this first hand because it happened to me in 1991 when I attended Special Forces Selection and Assessment Course. I thought I had prepared properly by conducting multiple foot marches, conditioning my feet and ensuring I had calluses. After moving through a swamp for over an hour, the calluses softened up and started to peel away from my feet. Over the next two weeks, my feet took a beating and bled every day. From that point on, I was conducting daily triage on my feet just to make it through the selection process. It was a very painful lesson to learn.*

This can happen fairly easy at LDAC especially if it is raining a lot or if you have sweat soaked boots that are not rotated regularly to allow drying. Problem areas are the ball of the foot, heel, and toes and especially small toes. For small toes, calluses can get so bad that the skin of the entire toe comes right off! OUCH, that hurts! Continuing to work through this issue requires perseverance, but will pay off in the end. The best prevention is to use a pumice stone. When using a pumice stone or callus reducer, ensure the feet are dry when filing down the rough skin. This should be done in stages because if you become too aggressive, your skin will become raw and irritated. In the beginning, callus removal is done every day until it is only necessary once a week to maintain your foot health. Once you go to weekly, you can apply lotion as often as you like. Next, moisturize the foot with a heavy-duty cream; Kiehl's makes an excellent product called Klaus Heidegger's All-Sport Foot Cream.

Maintaining long toenails is another mistake that can cause problems because if they are too long and the shoe is too small in the toe box, the toenail continually bumps against the end of the shoe, or the toenails catch on the top of the toe box. The remedy for this issue is quite simple; use toenail clippers on a regular basis followed by using a nail file to smooth the edges. Toenails should be cut straight across to avoid irregular growth, such as ingrown toenails. This will help avoid black toenails and blisters under nails.

**Leadership Insight 33:** Cadets must conduct foot marches to break in their boots and determine where blisters will form. At LDAC a Cadet walks approximately 75 km while conducting land navigation, foot marches, SSTX, movement to the FOB and PSTXs.

**COLD HARD FACT 19:** If Cadets don't complete the foot marches at LDAC they will receive negative spot reports from the Cadre, be ranked low on peer evaluations and be a liability to their platoon.

**COLD HARD FACT 20:** Cadets will carry rucksacks at LDAC that will have a minimum of 35 lbs in them.

*Blister and Abrasions Treatment* - Although there are several methods to treating blisters and abrasions, there are certain factors that influence the best type of treatment. The factors which influence the type of treatment are: whether you have medical assistance available, the size of the blister, the location of the blister, the condition in which treatment takes place, the type of activity you are involved in and the time available for treatment. While at LDAC, there are dedicated medics who make scheduled stops daily to inspect Cadets' feet and treat blisters if required. For some odd reason, Cadets don't tell the medics they have an issue because they are either afraid or don't want to place themselves in the limelight. Cadets must understand that the medics are there to assist them and ensure health conditions are being met. There may be times when a Cadet does not want to wait for the medic or the blister surfaces after seeing the medic or it could be the Cadet is stubborn or hard headed. Although my recommendation is to allow the medics to treat the blisters, I still believe that every Cadet should be capable of treating his or her own blisters and abrasions. Below is some guidance to how to treat the blister:

To treat blisters, carefully wash around the blister with soap and water, being careful to not break the skin if the blister is unbroken. Empty the blister, if unbroken, by pricking the lower edge with a needle or knifepoint, which has been held in a flame, however, an alcohol swab can be used to sterilize the knife or needle if available. Create a small drainage hole at the edge of the blister and BE CAREFUL not to cut the foot when using a knife because the skin can be tough! After draining the blister, do not remove the skin because it provides protection to the wound to avoid infection. Cover the blister with an absorbent adhesive bandage or similar dressing smoothly applied with adhesive plaster extending beyond its edges; an excellent product to use is Spenco New Skin. After applying the dressing, dust the outside and the entire foot with foot powder but make certain that not too much is used because the powder can cake and become irritating. Foot powder lessens friction on the skin and prevents the raw edges of the adhesive plaster from adhering to the socks. Check the blister periodically to see that it is drying properly.

Abrasions and cuts on the foot can be covered with an absorbent adhesive bandage, under which they usually heal rapidly. The adhesive plaster should be absolutely smooth, so it can serve as a second skin. New Skin is an excellent protective measure that is invisible and acts as a bandage.

**Leadership Insight 34:** The best blister prevention is to get out and toughen up your feet by conducting at least two foot marches of at least five miles per month. Don't be a liability to your platoon at LDAC. Prepare yourself!

**COLD HARD FACT 21:** Blister prevention is much better than blister treatment.

If you do not dedicate the time to prepare your feet properly before attending LDAC, you will need to learn how to treat blisters and carry a blister kit. At www.myleadersource.com, we offer one of the best blister kits out on the market. Our blister kit contains everything that you will need to ensure you're prepared to treat those annoying blisters so that you can continue to train without any issues. Our blister kit contains a family of products that combines Glacier Gel, Moleskin, antiseptic towelettes and alcohol wipes into an ultra light foot care kit.

*Other Health Factors* - You are allowed to take prescriptions and vitamins so pack a multi-vitamin to get the recommended daily requirements to maintain your health and help ensure the absorption of key ingredients. Cadets who fall out during training usually did not eat a balanced meal or became dehydrated by not drinking plenty of water; you must eat while at LDAC and more importantly you must eat healthy so stay away from the short order line and put at least two different colors of vegetables on the plate. Your body loses approximately .9 liters of water through sweating and breathing per hour during stringent activity so drink water on a continuous basis to remain hydrated for each day of training at LDAC. The barracks are a feeding ground for germs so open the windows before leaving for each day's training in order to allow air to flow through and lower the possibility of becoming sick.

**Leadership Insight 35:** Use the hand sanitizer and wash your hands at every opportunity to avoid germs. Don't be afraid to go on sick call if you need sanitizer or medicine. Go see the doctor for spider, chiggers and mosquito bites before they become infected; use bug spray because the mosquitoes are HUGE!

# Post-LDAC

## A Reality Check

# 13 Accessions Explained

Congratulations, you completed LDAC without a hitch! It is the ROTC Rights of Passage to becoming a Military Officer and the first of many major training events that you will complete during your military career; no matter how long or short. During the days at LDAC, you will mature, demonstrate the potential to become a leader and have some cool stories to tell the under classmates during your MS IV year. Prior to attending LDAC, you participated in multiple training events, took countless APFTs, performed numerous land navigation exercises and received an On Campus CER. Suppose during your attendance at Warrior Forge, you scored a 325 on the APFT, passed land navigation on the second attempt, gained an excellent performance rating after receiving a migration approval on the LDAC CER from your RTO. Now that you have conquered the Great Beast of LDAC, I am sure you have several questions. From my experience, the two most popular are: 1) *"How does LDAC calculate into my placement on the National Order of Merit List (OML)?"* and 2) *"What is my placement?"*

The Accessions Timeline can be a bit confusing but nevertheless, it will be a hot topic that spawns many questions that will not be answered until the final packet is submitted in the month of September. In reality, The Accessions Timeline begins the day you join ROTC but Cadets don't worry about it until their MS III year when they are scheduled to attend LDAC. During the timeline, the most important time period in this cycle starts when you submit your paperwork to the Human Resources Technician (HRT) to be aligned with a specific regiment at LDAC. Once the Battalion HRT inputs and validates your information into the Cadet Command Information Management System (CCIMS), the Accessions Cycle has begun and you are on your way to becoming a 2LT. During LDAC your results are uploaded into the CCIMS database immediately following each Warrior Forge evaluated event so Cadre can track your performance and more importantly, start the calculation of your overall position on the National OML. After receiving credit from LDAC, the first order of business during the fall semester is to finalize your accession packet for submission no later than September (about 80% of this packet is completed prior to the end of the spring semester). Once this packet is submitted, your total score is computed resulting in your Order of Merit Score (OMS). During the month of October, the National OML, Education Delay Board and Commission Review Board are conducted followed by the Component Selection and Branching Board in November. Toward the end of October and no earlier than December, the Branch Board results are released to your PMS and then the PMS informs the Cadet. Sometime during the following spring semester, you will receive a BOLC II and BOLC III date followed by orders.

**COLD HARD FACT 22:** Not every Cadet will be awarded an active duty commission and not every Cadet will get their first branch choice.

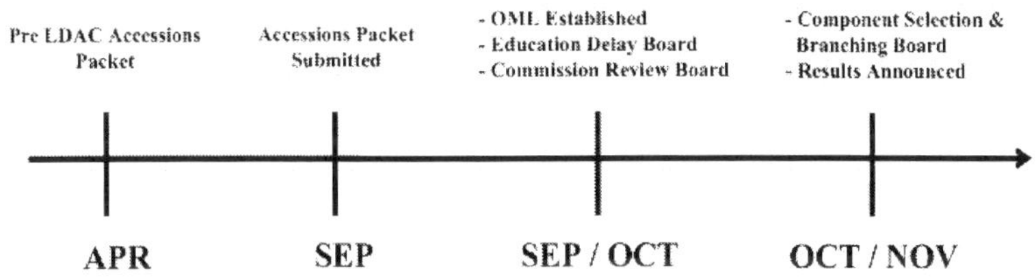


**Figure 20: Accessions Timeline**

**The National ROTC OML** is established by taking a Cadet's Academics, Leadership and Physical Abilities into account. Did you ever hear the acronym SAL? If you actually remember when you joined ROTC, you know that SAL stands for Student, Athlete and Leader which are the qualities ROTC consistently looks for when recruiting college students to join ROTC. The SAL directly corresponds to the three major categories used to evaluate a Cadet's position on the National OML which is a statistical model that standardizes and applies different weights to different events. The events in each of the categories add up to calculate each Cadets OML score which is based off of 100 points. The OML scores across the nation are very competitive and that is why this model goes 6 places past the decimal point to avoid any ties between Cadets. (e.g. 82.142349) Therefore, every fraction of a point can place you several positions higher.

The first category is *Academics* which accounts for a total of

40 points or 40% of the equation and directly relates to your Cumulative GPA from the college. Many Cadets and Cadre are confused on what final semester determines the GPA that goes forward to the Accessions Process. To clear up this uncertainty, the spring semester GPA is used! It becomes statistically harder to increase the GPA as a Cadet accumulates more hours over their college years.

**COLD HARD FACT 23:** The spring semester GPA before you are assessed is used to determine your Academics Score. For example, if your GPA at the end of the spring semester is 4.0 you get 40 points; if your GPA is 2.8, you get 28 points; fairly cut-n-dry!

The second and largest category is *Leadership* and accounts for 45 points of the OML Model. These 45 category points are balanced between **Leadership and PMS Experienced Based Observations.** Whatever you do, don't underestimate these two areas of the OML Model because they are HUGE!!!! There are 22 ½ points in each of these sections and they can make you or break you. These areas evaluate your true potential on how you lead and how you provide solutions to problems. The Leadership section under the Leadership Category accounts for your Warrior Forge Performance (E/S/N) during the 6 leadership positions to include your outcome in each of the 17 Leadership Performance Indicators, your Warrior Forge Overall Evaluation (E/S/N) and the first score on the Warrior Forge Land Navigation Test (Written/Day/Night). Each area listed accounts for a portion of the 22 ½ points and are based on an algorithm for each. If a Cadet receives an "E" for a position or dimensions, then they will receive 100% of the allotted points. If a Cadet receives an "S", then they only receive 85% of the allotted points and if a Cadet receives an "N", they only receive 70% of the allotted points within these categories. Table 10 shows the distribution of leadership points.

## Table 10: Leadership Distribution of Points

| LEADERSHIP EVENTS | POINTS |
|---|---|
| WARRIOR FORGE PERFORMANCE (E/S/N) – POSITIONS & DIMENSIONS | 6.75 |
| WARRIOR FORGE PLATOON TAC EVALUATIONS (E/S/N) | 11.25 |
| WARRIOR FORGE LAND NAVIGATION (1ST SCORE) | 4.50 |
| TOTAL LEADERSHIP POINTS | 22.50 |

The PMS Experienced Based Observations section under the Leadership Category accounts for the other 22 ½ points and includes the PMS MS III CER OML from your home campus, PMS Accessions OML (The MS III class ranking the fall semester after LDAC), PMS Accessions Potential Comments (Best Qualified, Fully Qualified, Qualified), Cadet Training / Extracurricular Activities and Language / Cultural Awareness. These events or areas account for the other 22 ½ points and rely heavily on your MS III year and LDAC performance. A small portion of these points are for Language / Cultural Awareness which can make an impact on the National OML. Table 11 shows the distribution of the PMS points.

## Table 11: PMS Experienced Base Observations

| PMS EXPERIENCED BASE OBSERVATIONS | POINTS |
|---|---|
| PMS MS III CER OML | 6.75 |
| PMS ACCESSIONS OML | 4.50 |
| PMS ACCESSIONS POTENTIAL COMMENTS | 4.50 |
| CADET TRAINING / EXTRACURRICULAR ACTIVITIES | 4.50 |
| LANGUAGE / CULTURAL AWARENESS | 2.25 |
| TOTAL PMS POINTS | 22.50 |

The **Cultural Understanding and Language Program (CULP)** was added into the ROTC OML Model in March 2009 after the Secretary of the Army emphasized an importance / requirement of CULP in institutional and operational Army training in the fall of 2006. Almost three years later, the decision was implemented by establishing the CULP program. The methodology was to establish an incentive to persuade Cadets to complete CULP activities to improve their Order of Merit Score (OMS). The new category was added to the Leadership portion of the ROTC OML Model but the total number of points did not increase. Instead the section Cadet Training / Extracurricular Activities was reduced from 6.75 to 4.50 and 2.25 points were added into a new category named Language / Cultural Awareness. See Table 11.

The new OML Model awards points to Cadets who declare their major in a foreign language, successfully complete language courses, complete Rosetta Stone Language Training or Study abroad in non-English speaking countries. The CULP only rewards Cadets points for languages that are strategic in nature to the needs of the Army. Strategic languages such as Spanish, French, Russian and Portuguese receive CULP points but they are not weighted the same as "critical strategic languages". The Department of the Army (DA) has declared and identified "critical strategic languages" that have strategic purpose and are critical to the Army. One key note is, if DA removes a language from the "critical strategic" list, the language remains weighted on the OMS credit for three years. It is very common that these languages change depending on the United States strategic plan. The CULP places a greater emphasis on these languages by weighting them with higher points. Although 201 points are available in the CULP, no Cadet can obtain all possible points because they cannot major in both a critical and non-critical language at the same time.. Table 14 identifies the critical strategic languages identified by DA.

## Table 12: Critical Languages

| CRITICAL STRATEGIC LANGUAGES | |
|---|---|
| ARABIC | CHINESE MANDARIN |
| PERSIAN DARI | PERSIAN FARSI |
| HAUSA | INDONESIAN |
| KOREAN | PASHTO |
| SWAHILI | URDU |

The CULP Activities are divided into four major categories with seven different activities. The *first category* is to major in a language. If a Cadet chooses to major in a critical strategic language they receive 15 points per year for up to three years for totaling 45 points. Whereas if a Cadet chooses to major in a non-critical language, they receive 10 points per year up to three years totaling 30 points. The *second category* is to successfully complete a foreign language course that was taken at a college. Although a foreign language from high school can be added to CCIMS, it does not count toward CULP points. Even though this category does not give a Cadet as many points as majoring in a language, there are still points. A total of 15 points per year can be awarded for passing a critical language course and 9 points per year for passing non-critical language courses; 5 points per course for critical strategic languages and 3 points per course for non-critical. The *third category* is completing a Rosetta Stone Course in either a critical or non-critical language. The maximum amount of points is 5 points per course for a critical language and 3 points per course for a non-critical language; limited to 1 course per year in each activity for a total of three course over a three year span. In my opinion, every Cadet should gain these points since Rosetta Stone is offered for free at Army Knowledge Online (AKO). I have taken several of the course and they are very easy to complete. The *fourth category* is to Study Abroad in a non-English speaking foreign country. Cadets must take a foreign college level course at a academic institution which must be depicted on the students college transcript; the course does not have to be a language course so go for the underwater basket weaving class and explore the culture. A total of 10 points are authorized per year for a total of 30 points over three years. Table 15 shows a detailed breakdown.

## Table 13: CULP Breakdown by Activity

| CULP ACTIVITY | POINTS | MAX PER YR | MAX POINTS |
|---|---|---|---|
| MAJOR IN A CRITICAL LANGUAGE (MUST BE DECLARED MAJOR AT TIME OF BOARD FOR 3RD YEAR CREDIT, 1ST AND 2ND YEAR MUST TAKE LANGUAGE COURSE IN EACH SEMESTER/QUARTER OF THAT YEAR) | 15 | 15 | 45 |
| MAJOR IN NON-CRITICAL LANGUAGE (MUST BE DECLARED MAJOR AT TIME OF BOARD FOR 3RD YEAR CREDIT, 1ST AND 2ND YEAR MUST TAKE LANGUAGE COURSE IN EACH SEMESTER/QUARTER OF THAT YEAR) | 10 | 10 | 30 |
| PASSING A CRITICAL LANGUAGE5 COURSE (MUST BE ON COLLEGE TRANSCRIPTS, HIGH SCHOOL DOES NOT COUNT) | 5 | 15 | 45 |
| PASSING A NON-CRITICAL LANGUAGE COURSE (MUST BE ON COLLEGE TRANSCRIPTS, HIGH SCHOOL DOES NOT COUNT) | 3 | 9 | 27 |
| ROSETTA STONE IN A CRITICAL LANGUAGE | 5 | 5 | 15 |
| ROSETTA STONE IN A NON-CRITICAL LANGUAGE | 3 | 3 | 9 |
| STUDY ABROAD IN A NON-ENGLISH SPEAKING FOREIGN COUNTRY (MUST BE COURSE ON TRANSCRIPT OF NON-US COLLEGE LEVEL ACADEMIC INSTITUTION, COMPLETED COURSES DO NOT HAVE TO BE LANGUAGE COURSES) | 10 | 10 | 30 |
| TOTAL POINTS AVAILABLE | | | 201 |

The most important piece of information that Cadets want to know is "What is the Impact?" with only 2.25 points at stake. The CULP Impact is two fold which has a positive affect for Cadets who take some form of foreign language training and a negative affect for Cadets without a foreign language. Cadets who put a higher weight on Cadet Training / Extra Curricular Activities will suffer the most because the category was reduced by 2.25 points to allow for CULP. Those Cadets that did not participate in the Training / Extra Curricular Activities and have foreign language training have the most to gain because the category is new and designed only for foreign languages. For example, a sensitivity test was conducted against the 2009 OML to determine how many Cadets would be affected. Below are the results:

- The largest increase in position was 475 places on the OML
- The largest decrease in position was 289 places on the OML
- A total of 17 Cadets exchanged places for selection as Top 10% in the nation
- A total of 31 Cadets exchanged places for selection as Distinguished Military Graduate (DMG)
- A total of 18 Cadets changed places for selection to Active Duty

The bottom line is that the CULP will have very little impact on Cadets overall OMS but the second and third order of effects can either increase or decrease a Cadet's position on the

National OML depending on their foreign language activities. If you are a Cadet that enjoys competing for the schools such as Airborne, Air Assault etc. or the type of Cadet that participates in extra-curricular activities shown in Table 12, you may consider to study a foreign language to remain competitive. I suggest reading the book "Who Moved My Cheese" by Spencer Johnson to understand the world of change. Buena Suerte!

The Cadet Training / Extra-Curricular Activities / Athletics captures other leadership opportunities outside of ROTC since a true leader is a leader outside of the military or ROTC. There are several Cadet Training Opportunities that offer an additional 5 points each. Cadet Troop Lead Training (CTLT) is an example of this type of training and is offered after completing LDAC as long as the Cadet passes all requirements without a retest. One key note, is that training opportunities must be available to all ROTC Cadets before additional points can be received. This rule exists so that prior enlisted soldiers do not have an advantage for schools or training that is only available while on active duty. These activities enable Cadets to earn additional points for each year they participate in the activity; the goal is to develop the whole person. If a Cadet participates in one of the listed activities they earn 5 or 10 points per year depending on the category. However, just as in life, there are exceptions to the point system mention and are depicted by an asterisks in the table. Some activities offer Cadets additional points due to their responsibilities and stringent requirements. Cadets must ensure they follow up on the documentation of these activities because Cadre may not be aware of Cadets extracurricular activities. The Extra-Curricular Activities/Athletics are listed below in Table 14:

## Table 14: Cadet Training and Extra-Curricular Chart

| ACTIVITY | YR1 PTS | YR2 PTS | YR3 PTS | TOTAL PTS |
|---|---|---|---|---|
| EXTRA-CURRICULAR ACTIVITIES | | | | |
| COLOR GUARD | 5 | 5 | 5 | 15 |
| DRILL TEAM | 5 | 5 | 5 | 15 |
| RANGER CHALLENGE | 5 | 5 | 5 | 15 |
| ROTC RECRUITER | 5 | 5 | 5 | 15 |
| * USAR/NG/SMP | 5 | 5 | 5 | 15 |
| * COMMUNITY SERVICE | 5 | 5 | 5 | 15 |
| STUDENT GOVERNMENT | 5 | 5 | 5 | 15 |
| BAND MEMBER | 5 | 5 | 5 | 15 |
| DEBATE TEAM | 5 | 5 | 5 | 15 |
| * ELECTED OFFICIAL OF ORGANIZATION | 10 | 10 | 10 | 30 |
| * LEADER (PRESIDENT/CAPTAIN) | 10 | 10 | 10 | 30 |
| * RESIDENT ADVISOR | 10 | 10 | 10 | 30 |
| PEER EDUCATOR OR TUTOR | 5 | 5 | 5 | 15 |
| * FULL-TIME JOB | 10 | 10 | 10 | 30 |
| * PART-TIME JOB | 5 | 5 | 5 | 15 |
| MAXIMUM ACTIVITY POINTS AVAILABLE | | | | 285 |

* More responsiblity and stringent activites

The final category is Physical which has a total of 15 points and includes Varsity, Intramural or Community Team Athletics (1.5 pts), On Campus Fall APFT (1.28 pts), On Campus Spring APFT (1.91 pts), the Cadets 1st APFT Score at Warrior Forge (9.56 pts; A HUGE IMPACT) and Swimming requirements (.75 pts). When you observe the points available for the Physical Category, the most points a Cadet can gain is from the Warrior Forge APFT. The remainder of the Physical points adds up to 5.44 but if you fail your APFT at LDAC, you will need as many of the points as you can get.

**COLD HARD FACT 24:** If a Cadet fails their first APFT, the maximum amount of points that can be obtained is the total score from the first APFT. Yes, this holds true even if the Cadet scores a 300 on the second attempt. The swim test and CWST will be conducted at your school and may be conducted at LDAC, if either event is failed, you will leave Warrior Forge without LDAC credit.

## Table 15: Athletics Scoring

| ACTIVITY | YR1 PTS | YR2 PTS | YR3 PTS | TOTAL PTS |
|---|---|---|---|---|
| ATHLETICS | | | | |
| * VARSITY ATHLETICS | 5 | 5 | 5 | 15 |
| INTRAMURAL TEAM | 5 | 5 | 5 | 15 |
| COMMUNITY ATHLETICS | 5 | 5 | 5 | 15 |
| MAXIMUM ATHLETICS POINTS AVAILABLE | | | | 45 |

The final category is *Physical* which has a total of 15 points ınd includes Varsity, Intramural or Community Team Athletics (1.5 ɔts), On Campus Fall APFT (1.69 pts), On Campus Spring APFT 2.36 pts) and the Cadets 1st APFT Score at Warrior Forge (9.45 pts; **A HUGE IMPACT**). The total points available for the Physical Category is 13.5 with most of these points obtained from the Warrior Forge APFT. The total Physical points on campus is 4.05 but if you ail your APFT at LDAC, you will need all of these points. So, lose the Tiny Heart Syndrome (THS) and do some PT! Of note, in FY 2012, he Army wants to add incentive points for those Cadets with degrees ı hard science, math and engineering degrees.

lease replace the first paragraph on page 270 with this updated data.

**COLD HARD FACT 25:** If you are in the Platoon Top Five you receive 1 additional point added to your final OMS. If you are a RECONDO, you will receive an additional ½ point added to your final OMS. Therefore, the maximum possible OMS is 101 ½.

Once the National OML is established, the next step is the *Branching Process* that determines which Cadets go on Active Duty and which ones do not. However, before the National OML roster can be determined, Army Nurses and other individuals selected in the Medical Specialty Corps, Reserve Forces Duty (GFRD and non-GFRD volunteering for Reserve Forces Duty) and Cadets who applied for Education Delay are removed. Once these personnel are removed, the remainder of the Cadets are eligible for active duty; however, active duty slots are not guaranteed because there are only a certain number of active duty slots determined by the Deputy Chief of Staff per year. Of course, each year officer strengths change due to the needs of the Army so those not selected for Active Duty are automatically referred to the RFD. The final step is to assign branches based off the Cadet's OML score and position, release AC/RC branching results and distribute the results. From this point, Cadets can receive their first choice, second choice or even their last choice. Even if you receive your last choice, you should realize that you have an incredible opportunity to succeed and lead soldiers that will make a difference in the world. One final note is that the Cadet's PMS must manage expectations and be honest with the Cadet after the Active Component and Reserve Components results are released.

Once the Accessions Packet is submitted, the waiting begins and you have no control over what happens, so you have to make sure you have war gamed how you will get your branch of choice. If you did your best and accomplished the majority of the tasks that were discussed prior to this point, you have an excellent chance to be in the top 10% of the National OML. However, if you did not prepare correctly and are considering begging your PMS for help to get your first branch choice, it is futile. Instead I recommend working out some specific Courses of Actions (COA) on how to negotiate your position before submitting the Accessions Packet. One COAs is to use the Officer Career Incentive Program to increase your odds in getting what you want. The purpose of the program is to retain quality officers by allowing Cadets to participate in all of the programs. However, a Cadet can only select two of three Officer Career Incentive Programs. If a Cadet participates in the program, they incur an additional three year Active Duty Service Obligation (ADSO) but the Cadets may receive their first branch even if they did not do so hot while in ROTC. Whatever the case, NEVER GIVE UP and always ask for what you want and follow up with due diligence.

**COLD HARD FACT 26:** Fail to use available Officer Career Incentive Programs to get the branch of your choice and you may receive a branch that you will not enjoy. The Active Component Branching Models fills branches by the following sequence: 1) Top 10% of the National OML receives their 1st branch choice; 2) Each branch then fills 50% of the available slots using the Cadet National OML. Since many of the branches do not have very many allocated slots for the year, it is very important for Cadets to war game if they will submit an ADSO; 3) Once the branch fills to 50% of the allocated slots for the year, Cadets that submitted an ADSO fill allocated slots until the branch is 65% full; 4) Once a branch fills 65% of its allocated slots, the remainder 35% of the allocated slots are filled by Cadets who are below the 50% line on the National OML; 5) Assign remaining Active Duty selected Cadets by needs of the Army.

There are a few **Rules of Thumb** for establishing the National OML. Although it does not affect most Cadets, the *Educational Delay Board* meets to address three specialty branches: Judge Advocate General Corps, Chaplain Corps and Army Medical Department (AMEDD) that encompasses Medical Specialist Corps, Medical Corps, Dental Corps and the Veterinary Corps. These branches require special consideration because they require additional years at college after LDAC before a Cadet can receive their commission.

The Active Component Basic Branch determines what branch Cadets receive and if they will be Branched Detailed to another branch for their first three years of active duty service. Believe it or not, every Cadet selected for Active Duty, except for the top 10% and those Cadets who receive their branch choice with an ADSO, can be chosen for the Branch Detail Program. The strategy behind the Branch Detail Program is that there are four branches that are considered donors and four branches that are considered recipients. The four branch donors are Signal Corps, Military Intelligence, Adjutant General and Finance but they can change from year to year if requirements change. These four branches fall under Sustainment and Force Protection Branches and do not require a large need for Officers at the Lieutenant rank. These branches loan out their Lieutenant Officers for approximately three years, not including Individual Entry Training, to serve in the Fire and Maneuver Branches. The Fire and Maneuver Branches are opposite of the Sustainment and Force Protection Branches in the respect they require a large amount of officers at the Lieutenant rank and less at the Captain rank. The four branches that are in the Fire and Maneuver element are Infantry, Armor, Field Artillery and Chemical Corps but they too can change from year to year if requirements change. During the initial branch selection, Cadets are chosen to fill positions at the Captain Level and then they select donors to be Branch Detailed. The Branch Detail Program is a great tool to assist Cadets in receiving their branch of choice since not all Cadets fare well from the OML Model due to GPA, LDAC performance or weak individual skills. A common area that Cadets try to avoid is the ***Zone of Death*** because no matter what you do, the likelihood of getting your desired branch is a very low percentage. There are many different strategies in trying to overcome the ***Zone of Death*** but the most common is by requesting Branch Detail on their Accessions Packet to assist in receiving their first branch of choice. Although a Cadet that is in the ***Zone of Death*** will more than likely get their second branch of choice (branch dependent), I can't stress enough to NEVER GIVE UP. A success story that comes to mind is a former Cadet that went to the University of Central Florida who was smack dab in the middle of the ***Zone of Death***. The Cadet had approximately a 3.1 GPA and received an Excellence Rating at LDAC these qualifications were not good enough to avoid the ***Zone of Death***. Although the Cadet placed Aviation as his first choice but because of his placement on the OML, he was counseled that it would be almost impossible to receive this branch. After the Accessions Process, he received his third choice with a Branch Detail submission. After talking with him for a while, a game plan was made to submit for a transfer to Medical Services Corps (MSC) and then apply to become a pilot for the MSC. Although it took a while, he was transferred to MSC and then selected to be a pilot due to his perseverance and desire to get what he wanted. Even if you don't get your branch of choice, don't give up because you can accomplish anything in the Army. There are always options but it is up to you to follow through with that option. I have known two former ROTC Cadets that put in a request to become a pilot and received it after serving a year on active duty in another branch. Like I said, NEVER GIVE UP because no one manages your career better than yourself!

**COLD HARD FACT 27:** You can still get the branch of your choice even though you fall in the ***Zone of Death*** by requesting a donor branch with a branch detail to the actual branch you desire as your second choice (e.g. Signal Corps as first with branch detail in Infantry). Aviation, Infantry, Armor,

Engineer, Military Police, Finance, and Military Intelligence usually fill to 65% with ADSO. Cadets that want these branches have a higher probability of falling into the ***Zone of Death***.

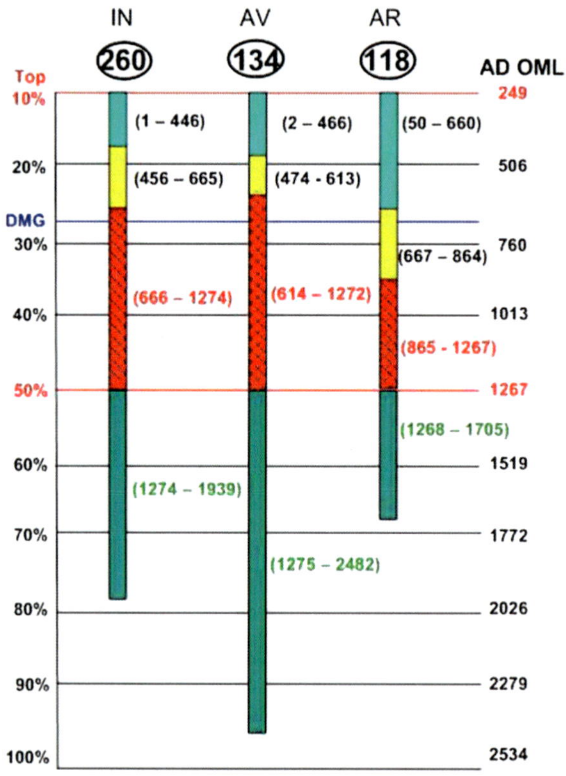


**Figure 21: Zone of Death**

The previous figure explains the ***Zone of Death*** in detail. During the fall semester of your MS IV year, your PMS will give you a class on the Accessions Process and it will include a slide similar to the previous figure. The slide will show all branches and more than likely take over half the class to explain because it can be very confusing to understand. Therefore, we will cover the layout of the figure and the process in detail to assist your learning. I suggest referencing the figure as we discuss it over the next few paragraphs.

The branch is shown at the top of the figure with the number of allocations for that year in the circle. As one can see, the Armor (AR) branch has 118 slots available this year. The lines and percentages on the left hand side are directly correlated with the numbers on the right hand side for the Active Duty (AD) OML. For example, the top 10% of the National OML is capped at the top 249 Cadets. The top 20% of the National OML, who will receive Distinguished Military Graduate (DMG) status, is calculated from the 4500 plus Cadets that will commission in that year. Once the Active Duty OML is determined, the DMG Cadets that will be serving in the National Guard or Reserve are removed from the list resulting in approximately 2500 Cadets to be place in the branching model. Once these numbers are removed, the DMG line appears to have shifted from the top 20% to almost 30%. However, this is not the case because the DMG list was based off the original 4500 plus Cadets. The light blue box represents the first 50% of the total allocated slots filled by Cadets in that branch, the yellow box represents the percentage of slots filled by Cadets with an ADSO (up to 65% of the branch) and the green box represents the remainder percentage of the branch that was filled by Cadets who were below the 50% AD OML Line. The red box represents Cadets that fall in the ***Zone of Death!*** The numbers in parenthesis along the side of the colored boxes represent the start and end of the National OML which filled that group.

For example, in the AR Branch, there were 118 allotted slots for Cadets this year. Therefore, the first 59 Cadets by OML filled up the first 50% (59 slots) of the allocation that asked for AR. If any of these Cadets submitted an ADSO to received AR, the ADSO was not charged to the Cadet because they received the branch on their own merit. Cadets with AD OML ranking 50 – 660 who asked for Armor, made up the first 50% of the Branch. The next 15% of the branch allocation was filled by Cadets who requested Armor Branch with an ADSO. Since there are only 18 slots available for ADSO (50% - 65%), these slots were filled by Cadets in the top half of the AD OML between 667 -864. The remainder 35% of allocated Armor slots went to Cadets who fell in the bottom half of the AD OML (1268-1705) that asked for Armor with an ADSO. Now for the alligator tears! Cadets that find themselves in the top 50% of the AD OML who requested Armor but did not get branched Armor within the first 65% of the allocated slots (77 of 118) fall in the ***Zone of Death*** and will be automatically sent to their second branch choice. This process is then repeated and if the Cadet falls in a second ***Zone of Death*** for their second choice, they are moved to their third choice and the process is conducted again. There have been DMG Cadets who selected their top three branch choices as AV, AR and IN who was given their fourth branch choice of MI.

So, what should a Cadet take from this long exhausting process? War Game with your PMS and understand where you fall within the AD OML, what the likelihood is that you will get your first branch choice from your placement and if you should engage in an incentive program to assist you in getting your branch of choice.

Below is a sample timeline for the Accessions Process but don't lock these dates in stone because they will change or fluctuate each year:

| Date | Event |
| --- | --- |
| Spring Semester | Battalions Input & Validate data in CCIMS |
| 18 August | End of LDAC Commission Cadets data into CCIMS |
| 15 Aug – 3 Oct | BDEs review & validate accessions / HQCC reviews accessions CCIMs data |
| 4 Sep | End of LDAC Commission Cadets Standby Board |
| 8 Sep | Last day to enter all OML data into CCIMS & check OML packets |
| 15 Sep | Release Branching Results for End of LDAC Commission Cadets |

| | |
|---|---|
| 17 – 19 Sep | Commission Review Board |
| 22 Sep | Publish National OML, RD & AD Selection, DMG and top 10% |
| 5 Oct | Last date to adjust Branch & ADSO Choices; Cadet & PMS comments complete |
| 7 – 9 Oct | Educational Delay Board |
| 14 – 17 Oct | AD/RD Branching Boards |
| 23 Oct | Release Educational Delay, AD & RD results & begin publishing orders for Dec grads |
| 17 Nov | Last date to request Post for ADSO |
| 24 Nov | RD Cadets without a BSR assigned a branch based on the needs of the Army. |

## War Gaming is a Work of Art!

Each branch has a specific number of positions that are available for United States Military Academy (USMA), Reserve Officer Training Corps (ROTC) and Officer Candidate School (OCS) Commissioned Officers. These numbers are divided so that all three commissioning programs receive a portion of the slots available with OCS making up the shortfalls. Of course the intent is that the top 10% of each program get their branch of choice and from that point on it is determined by their Order of Merit Score. Not all ROTC Cadets do well with the ROTC OML Model so there are three programs that increase a Cadet's Active Duty Service Obligation from four to seven years as a Scholarship Cadet or three to six years as a Non-Scholarship Cadet to assist them in getting their branch of choice, additional schooling or their desired location to be stationed.

The Branch for Service Option is the most popular and simplest to request. By selecting this option, you are telling the Army that you are willing to do an additional three years of active duty to get your branch of choice. If there are more Cadets requesting that branch than there are positions, then the OML is used to determine who gets selected. Other factors considered are manpower requirements, gender and medical requirements, etc. Cadets that do not choose the Branch for Service Obligation do not decrease their odds of receiving their first branch of choice due to changes that have been made to the DA Selection and Branching Board Process. If a Cadet is placed in the top 10% of the National OML, the Branch for Service Obligation is ignored and the Cadet does not incur three additional active duty years. There are several branches that can fill up 50% of the positions before the Distinguished Military Graduates (Top 20%) are assigned. These branches are Aviation, Infantry, Armor, Military Intelligence, Finance and Medical Services. Finally, a Cadet can submit up to two branches that they are willing to incur an additional three years of active duty for using this program. Cadets can also request an dual

ADSO for two different programs.

The Graduate School for Service Option is a great way for a Cadet to secure an opportunity to attend a full time, fully-funded Graduate Program within the United States while receiving full military base pay and entitlements at the current rank. During your attendance at school, you are still subject to UCMJ, required to maintain height/weight and personal appearance and take an APFT every six months. If you sign up for the Graduate School for Service Option, it does not mean you have to execute the option; however, if you do choose to exercise this option, there is an additional three years of active duty service to include an Advanced Civil Schooling (ACS) payback of an additional three months for each one month in graduate school. Depending on when the option is exercised, it makes a difference on how much time you owe Uncle Sam. If you exercise the option after your six year of service then you will have already met the three year ADSO and will only have the ACS payback remaining. The ACS payback is a minimum of three years after completing the schooling. Officers can attend the program early under the Graduate School for Service Obligation but the contract specifies that it be between the sixth and tenth year of active commissioned service. As with anything in the Army, you will have to work with your Assignments Officer to determine the specific date but it is your responsibility to secure admission to a school from a list of DA approved schools. The Army and the college you choose both use your GPA or GRE/GMAT score as criteria to be selected for this program. One key note is that if you receive a Graduate Degree as a Cadet you are not allowed to participate in this program to obtain a second graduate degree. Finally, if you participate in this program, you can apply to participate in the FLEP program, medical school program, or one of the scholarship programs and then your Graduate School for Service contract will be voided.

**COLD HARD FACT 28:** Cadets that did not do so well in school and do not have a high GPA are limited to what colleges will accept you as a student.

**Leadership Insight 36:** Take the GRE/GMAT immediately following graduation. You will be more prepared to take it now than you will in three years. If you bomb it, you can take it again in six months. Nevertheless, I would not accept no for an answer because many universities and colleges are very patriotic and can be persuaded to accept you as a student. Ask for a waiver!

The ACS Program is very different than the Graduate School for ADSO. The selection for ACS is very competitive and takes into account your GPA, GRE/GMAT, your military service record and your performance. Additionally the Army imposes much more restrictive parameters on the type of program, when you attend and the school you attend if you are selected in the ACS Program. After completion of the school, you usually have to serve a three year utilization tour in a functional area or as a West Point Instructor. Your entire service obligation is after you complete school which adds up to be six years. If Cadets don't sign up for the Graduate School for Service Option, during accesssions, once you are in the Army, you can pursue Graduate Degrees on your own time with your own money by going to school at night while at your current duty location. In addtion, some branches allow you to participate in an established Graduate Degree Programs while at the Captain Career Course. For example, both the Military Police and Engineer Branch offer such programs. The Engineer School offers a Joint Masters Degree in Engineering Management through the University of Missouri-Rolla. Participants must pay for the school out of their pocket but the Army offers Tuition Assistance. Furthermore if you receive a Graduate Degree from one of these programs or on your own at night, you are still eligible for ACS, the Funded Legal Education Program (FLEP) or the Graduate School for Service Option as long as the degree is not the same as the degree you earned on your own time using your own funding. Be aware that if you pay for a graduate program and attend on your own time, the Army does not reimbuse your monetary loss if you are deployed. Therefore ensure you keep your commander informed before making a decision that may not be supported due to upcoming missions.

The Post for Service Obligation is the last of the programs to assist ROTC Cadets. A Cadet can ask to be assigned to any specific location but the assignment may be limited to the number of 2LT available slots. There is also the fact that the

Cadet's branch may not be at that post. For example, if you are an Armor Officer, you will more than likely not be assigned to Fort Sam Houston, Texas. Since the assignment of choice is not a guarantee, you will not incur the ADSO if you are not given your choice or if your assignment is changed. Two other things to consider are 1) if you do receive your assignment of choice and you are deployed away from your post, the time will count against your time at that post and 2) participation in this program only covers your first assignment after initial basic skill training and is intended to allow you to serve one complete life cycle at that post.

War Gaming assist Cadets in receiving their desires during the Accessions Process but does not come with a guarantee. Participation in these service options is not mandatory and only Cadets selected to Active Duty are eligible to participate to include Active Duty Green-to-Gold and Early Commissioning Program (ECP) Cadets.

*Some Limitations and Final Thoughts*

1) Nurse Cadets are only eligible for the Post for Service Program and Aviation Cadets cannot participate in the Post for Service Program.

2) If an Aviation Cadet submits a Branch for Service Obligation, the additional three years are added to the Aviation Training Service Obligation.

3) If a Cadet submits a Branch for Service Obligation and does not volunteer for a Branch Detail, the Cadet will not be consid-

ered for Branch Detail if the Branch for Service Option is used.

4) If the Cadet volunteered for Branch Detail, then the Cadet will be considered for Branch Detail to include the Branch for Service Option.

5) If a Cadet submits an ADSO request and receives their branch choice on their own merit, the ADSO is not charged to the Cadet. Good Luck!

# 14 That's It!

The past thirteen chapters have covered the **COLD HARD FACTS** and **LEADERSHIP INSIGHTS** of what a Cadet should expect at LDAC and information about Accessions. Therefore this chapter is going to be short and sweet without a lot of recap because you either got it or you didn't. My intent when writing this book was to answer the majority of Cadets' questions prior to their attendance at Warrior Forge even though they might not know what those questions were before entering their MS III year. During my tenure instructing at UCF for three years, the questions were almost always the same, even the time of the year they surface. So by reading this book, you now have the answers to most of your questions before you were ready to ask them. For the Cadets who have read this book, the next step is to apply the techniques in training and during practical exercises that you experience at ROTC prior to attending LDAC. Now, it is up to you to apply this information so that it works to your advantage! If you are a true leader, then take charge and organize a study group at your campus to discuss these Leadership Insights which will create additional questions and discussions. You must be a leader by closing the loop and connecting the insights at a larger concept in order to be successful. You must conduct additional training outside of ROTC and volunteer for additional leadership opportunities so that you can apply these tidbits of information before you will understand the full scope of what has been discussed. Expect that you will make mistakes while applying learned techniques in ROTC but it is not expected that you make the same mistake twice or a third time. The best advice I can give Cadets reference ROTC and LDAC is to refer back to this book from time to time and revisit the information presented, ask questions during and after training and get involved with your peers to learn more insights and lessons learned. Finally, I suggest incorporating additional training after you complete the spring semester so when you go to LDAC you are current with Land Navigation, Physical Fitness, SSTX and Garrison Operations at a minimum. Many Cadets are not able to replicate this training because of the number of Cadets required to make a squad or patrol, the logistical requirements, the availability of Cadre to conduct evaluations and a sufficient training area. While I was at the University of Central Florida (UCF), I developed an additional training program after the spring semester called "School of the Cadet" that focused on the above mentioned areas to afford Cadets the option to practice and place finishing touches on their abilities before attending LDAC. This training was a major success in UCF accomplishments at LDAC and is what I believe necessary for Cadets to arrive at LDAC with the skills and confidence necessary to succeed.

I have included a Physical Fitness Plan to ensure you are

prepared for that first critical and emotional event at Warrior Forge since most Cadets slack off on this crucial event before attending. Once receiving the On Campus CER, if you find yourself not prepared to attend LDAC you still may have some options but it will cost you time, money and dedication. The first option is to reread this book as many times as you can before leaving for LDAC to include dedicating the time to learn necessary task during the MS III year. The second option is to ask Cadre to schedule additional training events during the summer so that you can prepare but this almost never happens because Cadre are taking leave, departing for Warrior Forge or The Leadership Training Course (LTC). Therefore, I offer a one week LDAC Prep Training Program for Cadets who want to increase their chances of raising their ratings from an "N" to an "S" or increase their rating from an "S" to an "E." The training I offer is tailored for Cadets who want to ensure they are fully prepared before attending LDAC, want to increase their performance, failed to properly prepare for LDAC during the MS III year or those who want to reinforce basic leadership skills before departing for LDAC. If you are interested in attending a 1-week training event after the spring semester before attending LDAC, **visit www.myleadersource.com** for the complete details. Finally, I leave you with the two remaining Cold Hard Facts about LDAC:

**COLD HARD FACT 29:** Hope is not a method for birth control or Army Operations, so if you fail to plan, plan to fail!

**COLD HARD FACT 30:** Just reading this book will not get you an "E"!

# Workouts

## 4-Week Running Program

This section is nothing more than a guide! If you are having a hard time passing your run, then this section is for you to train and become a better officer. If you are passing the run with 70 points or more, then subtract 30 seconds from the pace and do the same workout at a faster pace. If you are looking to max the run portion of the APFT, then subtract 1 minute from the pace and get ready to excel. This workout has worked for me in the past in developing Cadets, so take it to heart that you have to run and you have to have a regimented routine in order to succeed. Finally, this physical fitness plan is a guide and gives a great starting point to improve. If the workout is too easy for you, then adjust the workout and make another version for the "Freaks of Nature"; GOOD LUCK!

### Week 1

**Day 1** (Monday) - Run 3 miles @ 8 min pace or better

**Day 2** (Wednesday) - Intervals (Track)
2 warm up laps @ 75% of max speed
(2 min 30 sec per lap)
6 repetitions of fast / slow:
½ lap fast (80% of max) shoot for 45 sec to 1 min
½ lap slow (very slow shuffle)
2 cool down laps @ 75% of max speed
(2 min 30 sec per lap

**Day 3** (Friday) - Run 4 miles @ 8 min pace or better

## Week 2

**Day 4** (Monday) – Run 4 miles @ 8 min pace or better

**Day 5** (Wednesday) – Intervals (Run for 30 minutes)

First 2 min @ 75%

Run 1 min @ 90% of max (fast interval)

Run 2 min @ 75% of max (slow interval)

Run 1 min @ 90% of max (fast interval)

Run 2 min @ 90% of max (slow interval)

Continue sequence until 15 min; turn around and continue the sequence until 27 min

Finish last 3 minutes @ 60% of ma

Day 6 (Friday) – Run 5 miles @ 8 min pace or better

## Week 3

**Day 7** (Monday) – Run 4 miles @ 7 min 30 sec pace or better

**Day 8** (Wednesday) – Intervals (Run for 30 minutes)

First 2 min @ 75%

Run 1 min @ 90% of max (fast interval)

Run 2 min @ 75% of max (slow interval)

Run 1 min @ 90% of max (fast interval)

Run 2 min @ 90% of max (slow interval)

Continue sequence until 15 min; turn around and continue the sequence until 27 min

Finish last 3 minutes @ 60% of max

**Day 9** (Friday) – Run 5 miles @ 8 min pace or better

## Week 4

**Day 10** (Monday) – Run 3 miles @ 8 min pace or better

**Day 11** (Wednesday) – Intervals (Run for 30 minutes)

First 2 min @ 75%

Run 1 min @ 90% of max (fast interval)

Run 2 min @ 75% of max (slow interval)

Run 1 min @ 90% of max (fast interval)

Run 2 min @ 90% of max (slow interval)

Continue sequence until 15 min; turn around and continue the sequence until 27 min

Finish last 3 minutes @ 60% of max

**Day 12** (Friday) – Run 2 miles as you would on an APFT

## Week 5

**Day 13** (Monday) – Run 3 miles @ 8 min pace or better

**Day 14** (Wednesday) – Intervals (Track)

2 warm up laps @ 75% of max speed (2 min 30 sec per lap)

6 repetitions of fast / slow:

½ lap fast (80% of max) shoot for 45 sec to 1 min

½ lap slow (very slow shuffle)

2 cool down laps @ 75% of max speed (2 min 30 sec per lap)

**Day 15** (Friday) – Run 4 miles @ 8 min pace or better

* *Incorporate riding a bike for 30 minutes on Tuesdays throughout program*

* *Incorporate weight training (total body on machines) on Thursdays throughout program*

## Upper Body Round Robin (UBRR) WORKOUT

**Day 1** Chest- Bench Press 10 sets of 10 reps (same weight)
1 minute of rest between sets
Triceps- Close grip bench 3 or 4 sets of 10 reps
Biceps- Standing curls 3 or 4 sets of 10 reps
Abs (Work Out 1)

**Day 2** Back- Pull ups 10 sets of half your max
Shoulders- Military press 3 or 4 sets of 10 reps
Legs- Squats or Leg press 10 sets of 10 reps
Abs (Work Out 2)

**Day 3** Chest- Push ups 10 sets of half your max.
1 minute rest between sets

Triceps- Dips 10 sets of half your max.
1 minute rest between sets
Biceps- One arm dumbbell curls
3 or 4 sets of 10 reps
Abs (Workout 1)

**Day 4** Back- Heavy bent rows 5 sets of 5-10 reps
Back/Legs- Dead lifts 5-10 sets of 10 reps
Shoulders- Front, side, rear raises 3 sets of 10 reps
Legs- Leg curls and leg extensions 4 sets of 10 reps
Abs (Workout 2)

## Abdominal Plan

(Workout # 1)

Do 10 repetitions per exercise and do 2 sets. Stretch abdominals between sets.

**Raise Ups** – Lay flat on back with arms down by sides. Using only the abdominal muscles, raise the torso in an upward motion while maintaining the arms along your sides as if you were trying to row a boat.

**Standard Crunch** – Keep knees bent 90 degrees, fingers on temples, raise the torso using the abdominals and hold at the top of the repetition for 1 sec and control when lowering.

**Lateral Leg Raises** – Lay flat on back with arms at the shoulder level; legs are straight up in the air. Swing the legs side to side slowly while maintaining your arms out to your sides.

**Leg Raises** – Lay flat on your back with fingers on your temples. Raise legs up to 75 degrees and then lower them.

**Toe Raises** – Start out flat on your back with fingers on your temples and bring your legs upward to the start position as if you were hanging upside down from the ceiling. Then raise your feet as if you were attempting to touch the toes to the ceiling and hold it for 1 sec.

**Raised Leg Toe Touches** – Lay flat on back with legs raised in the air and arms elevated and side by side. Perform the exercise as if you were touching the toes. Raise up and hold for 1 sec while trying to touch your toes.

**Alternating Raised Leg Toe Touches** – Same as above except you alternate with each hand.

**Raised Leg Crunch** – Legs straight up in the air with your fingers interlaced as a fist. Attempt to touch your feet with the single fist formed by both hands and hold it for 1 sec at the top of the repetition.

**Lateral Raised Leg Crunch** – Lay on back with knees bent at 90 degrees, interlace hands together forming a fist and then raise up to the left side and hold for 1 sec then go to the right side doing the same maneuver.

**Reverse Crunch** – Lay flat on back with fingers on temple, pull legs into chest and squeeze for 1 sec.

**Single Leg Crunch** – Lay flat on back with fingers on temples; keep one knee bent at a 90 degree angle while the other leg is extended straight out. Perform the exercise by extending a leg up to your chest while doing a crunch at the same time. Squeeze for 1 sec when the knee and abdominals are compressed.

**Leg Extensions** – Sit on your butt with arms behind you or in front of you while maintaining your legs raised in the air approximately 6” off the ground. Extend your legs out to full extension and then bring them back into your gut.

(Workout # 2)

**Do 10 repetitions and 3 sets of each**

Hanging Crunches – Use a pull up bar to conduct these exercises by hanging freely. Raise the legs up into chest and hold for 1 sec. The 2nd set is hanging freely and raise your legs up to an L position. The 3rd and 4th set is hanging freely and bringing legs at an oblique angle to work the sides.

Lateral bends with weight – Obtain a 25 lb dumbbell or desired weight. Place the fingers of the free hand on your temple, and then pull that hand toward your side, which will raise the weight on the other side.

Crunch with weights – Place desired weight on the cable machine and use a rope to accomplish the exercise. Start by facing the machine on your knees and then conduct the exercise by placing the rope behind the neck and pulling the weight toward the floor while crunching the abs.

## The Cold Hard Facts

**COLD HARD FACT 1:** There is no secret warehouse that has all the knowledge dedicated to Cadets before they attend LDAC.

**COLD HARD FACT 2:** Cadets go to LDAC with the knowledge and insights that they have obtained since being affiliated with Army ROTC.

**COLD HARD FACT 3:** The LDP works and is a proven system!

**COLD HARD FACT 4:** A Cadet can receive an "N" in Leadership Performance Indicator (LPI) dimension on a Blue/Green Card and still receive an overall "E."

**COLD HARD FACT 5:** Cadets who do not receive LDAC credit can still be commissioned if they return the following year and pass; however, I would not count on it!

**COLD HARD FACT 6:** Failure to make a decision is worse than making a wrong decision. Therefore, if a wrong decision is made, correct the mistake and Charlie Mike (Continue the Mission).

**COLD HARD FACT 7:** Failure to do PT after completing the spring semester will deteriorate your physical fitness and you will FAIL the APFT. Fail the APFT twice and you are going home without LDAC credit.

**COLD HARD FACT 8:** If you fail an APFT event, you are an automatic "N" at LDAC. The APFT is a MUST pass to get LDAC credit. Remember the National Order of Merit List – Higher Score = Branch Placement.

**COLD HARD FACT 9:** Failure to mark the scorecard in the proper location with the clacker and document the point number or letter results in not receiving credit for that point. A lost scorecard results in a BIG FAT "NO GO" and a retest!

**COLD HARD FACT 10:** Cadets fail land navigation at LDAC for one of several reasons: 1) They do not keep an accurate pace count resulting in getting lost. 2) They are over confident and don't use their compasses. 3) They drift to the right or left because they stare at their compasses or continually go around obstacles in the same direction. 4) They leave their light on while moving from point to point which degrades their night vision capability. 5) They are terrified of the dark!

**COLD HARD FACT 11:** Get caught with prohibited items after the in-processing inspection and you may be booted from LDAC for an honor violation. Fail the urinalysis at LDAC and you will be booted from Army ROTC.

**COLD HARD FACT 12:** If a ROTC Cadre from your home campus makes a visit to you at LDAC, it is because your performance is lacking and you are considered an "At Risk" Cadet.

**COLD HARD FACT 13:** Failure to get an accurate count of platoon personnel, weapons and equipment when loading and unloading vehicles will result in an emotional event if a Cadet is left behind or the movement is delayed. This is a prime time for Cadre Override!

**COLD HARD FACT 14:** If you do not follow the one-third / two-thirds time rule and the 5 paragraph OPORD format, you will receive an "N" in the Domain Knowlege (DK).

**COLD HARD FACT 15:** Failure to adhere to the one-third / two-thirds rule, failure to use the 5 paragraph OPORD format, failure to use graphics or deliver an unobserved OPORD during TLP will result in the Cadet receiving an "N" in the Domain Knowledge (DK) indicator at a minimum. Failure to have a Cadre present while delivering your OPORD at any level results in an unobserved OPORD and a BIG FAT "N" in those multiple indicators.

**COLD HARD FACT 16:** A "Failure to Train" counseling statement results in a removal from Warrior Forge without LDAC credit.

**COLD HARD FACT 17:** If you miss movement, you will experience an emotional event called "Cadre Override" and it will not be fun! The result will be a negative mark on your leadership evaluation.

**COLD HARD FACT 18:** If you get blisters or an injury at LDAC, the Cadre and fellow Cadets are **NOT** going to feel sorry for you! They expect that you pull your share of the weight and drive on without whining or trying to get over by using the injury as an excuse. Evaluators are **NOT** going to treat you any differently or give you a break because you are injured. If your injury is not severe enough for a medical recycle or a return trip to home campus then you are fit for training and will be evaluated as if you were at 100%.

**COLD HARD FACT 19:** If Cadets don't complete the foot marches at LDAC they will receive negative spot reports from the Cadre, be ranked low on peer evaluations and be a liability to their platoon.

**COLD HARD FACT 20:** Cadets will carry rucksacks at LDAC that will have a minimum of 35 lbs in them.

**COLD HARD FACT 21:** Blister prevention is much better than blister treatment.

**COLD HARD FACT 22:** Not every Cadet will be awarded an active duty commission and not every Cadet will get their first branch choice.

**COLD HARD FACT 23:** The spring semester GPA before you are assessed is used to determine your Academics Score. For example, if your GPA at the end of the spring semester is 4.0 you get 40 points; if your GPA is 2.8, you get 28 points; fairly cut-n-dry!

**COLD HARD FACT 24:** If a Cadet fails their first APFT, the maximum amount of points that can be obtained is the total score from the first APFT. Yes, this holds true even if the Cadet scores a 300 on the second attempt. The swim test and CWST will be conducted at your school and may be conducted at LDAC, if either event is failed, you will leave Warrior Forge without LDAC credit.

**COLD HARD FACT 25:** If you are in the Platoon Top Five you receive 1 additional point added to your final OMS. If you are a RECONDO, you will receive an additional ½ point added to your final OMS. Therefore, the maximum possible OMS is 101 ½.

**COLD HARD FACT 26:** Fail to use available Officer Career Incentive Programs to get the branch of your choice and you may receive a branch that you will not enjoy. The Active Component Branching Models fills branches by the following sequence: 1) Top 10% of the National OML receives their 1st branch choice; 2) Each branch then fills 50% of the available slots using the Cadet National OML. Since many of the branches do not have very many allocated slots for the year, it is very important for Cadets to war game if they will submit an ADSO; 3) Once the branch fills to 50% of the allocated slots for the year, Cadets that submitted an ADSO fill allocated slots until the branch is 65% full; 4) Once a branch fills 65% of its allocated slots, the remainder 35% of the allocated slots are filled by Cadets who are below the 50% line on the National OML; 5) Assign remaining Active Duty selected Cadets by needs of the Army.

**COLD HARD FACT 27:** You can still get the branch of your choice even though you fall in the ***Zone of Death*** by requesting a donor branch with a branch detail to the actual branch you desire as your second choice (e.g. Signal Corps as first with branch detail in Infantry). Aviation, Infantry, Armor, Engineer, Military Police, Finance, and Military Intelligence usually fill to 65% with ADSO. Cadets that want these branches have a higher probability of falling into the ***Zone of Death.***

**COLD HARD FACT 28:** Cadets that did not do so well in school and do not have a high GPA are limited to what colleges will accept you as a student.

**COLD HARD FACT 29:** Hope is not a method for birth control or Army Operations, so if you fail to plan, plan to fail!

**COLD HARD FACT 30:** Just reading this book will not get you an "E"!

## Leadership Insights

**Leadership Insight 1:** The Officer Evaluation Report (OER) is an annual evaluation that all Military Officers receive and it is based off similar leadership behaviors or Leadership Performance Indicators (LPI) used to evaluate Cadets at LDAC.

**Leadership Insight 2:** When in charge, **BE IN CHARGE!** How many times have you heard that and asked yourself "What does that mean?" Here are some guidelines for taking charge:

1) Don't let another Cadet make decisions for you.

2) Use subordinate leaders and give them guidance and intent to accomplish the mission.

3) Be proactive and don't wait for tasks to be directed; act in accordance with commander's guidance and intent and refer back to Rule # 2.

4) If there is a lack of commander's guidance, seek guidance; make recommendations so you control the decision, then refer back to Rule # 2.

5) Lead by example; peers and subordinates will emulate what you do; actions are louder than words.

6) Show a genuine concern for your peers when in charge, do what is best for the squad, platoon, company, etc. and don't lead for a grade.

7) Control the situation! Make a decision even if it is the wrong decision and accept responsibility.

**Leadership Insight 3:** Use the STAR, TAR, TAR format; write accurately and legibly using bullet format or sentence form to ensure that you are short, clear and concise.

**Leadership Insight 4:** Dedicate time to studying ROTC information by performing daily physical fitness and keeping a positive attitude. Continue to strive for excellence even if you fail an event. You may be able to recover if you keep a good attitude and apply yourself.

**Leadership Insight 5:** Always strive to do your best, develop the less skilled Cadets in the platoon, be a team player and be a leader. But most of all don't have THS (Tiny Heart Syndrome).

**Leadership Insight 6:** If your goal is to get 60 points in each event, **YOU WILL FAIL!** Always do your best in each event during every APFT. Don't let your enemy, Tiny Heart Syndrome, win the battle!

**Leadership Insight 7:** Don't allow the APFT at LDAC to become an emotional event and whatever you do, don't wait to start training for LDAC one week before you leave; my recommendation is to have a continuous physical fitness program throughout the year but if that is not possible, start at a minimum of four weeks before departing for LDAC. Your preparation begins during your first physical fitness workout at ROTC with a more focused approach and higher intensity level during the MS III year; form and correctness must not be compromised! Attend every PT session and add additional physical fitness sessions outside of ROTC. The best way to prepare mentally and physically before arriving at LDAC is to have a plan for how you will train and get the most out of your performance. Run with someone faster than you for motivation and moral support.

**Leadership Insight 8:** Fail one land navigation event and the best Overall Rating you can obtain is an "N." Although your PTO can migrate you to a higher mark, The Warrior Forge Commander views land navigation as an "Individual Capabilities Test" and doesn't see the need to migrate Cadets.

**Leadership Insight 9:** When taking the Land Navigation Written Test, go through and answer the easiest questions first. Leave the more difficult questions (e.g. resection & intersection, road distance) to the end so if time becomes an issue, you can still receive a passing score. Ensure that you mark the correct question with the correct answer and never leave a question blank. If you are unsure of the answer, make an educated guess even if time is running out. Most of all remember, Hidden (Hill) – Valley – Ranch (Ridge) – Salad (Saddle) – Dressing (Depression) for the major terrain features.

**Leadership Insight 10:** Established water points are the only points that have chemical lights attached to them. These sites are known points and can be used as an attack point to assist in finding your point. However, talking to other Cadets is not tolerated and Cadre is positioned at these locations to ensure cheating is not occurring. Cheating results in a "Go HOME" score.

**Leadership Insight 11:** TACs very rarely check the inside of your hat, boots, or cargo pockets of the ACUs. (Hint, hint!).

**Leadership Insight 12:** Conduct an inspection of your equipment before leaving each point on the course after marking your scorecard with the clacker and appropriate letter or number.

**Leadership Insight 13:** All the points at LDAC are approximately 400 meters from an attack point and were emplaced using a GPS for accuracy.

**Leadership Insight 14:** Take approximately 20 – 30 minutes to plot your entire route before moving onto the course. Ensure you transfer the data to the Route Planning Reference Sheet and then go find those points.

**Leadership Insight 15:** Some Cadre is not aware of all the prohibited items, which result in some items not being taken during inspection.

**Leadership Insight 16:** Example of a Daily Activities Garrison

| ACTIVITY | TIME | REMARKS |
|---|---|---|
| WAKE UP | 0500 | VARIES DEPENDING ON TRAINING |
| FIRST CALL | 0530 | VARIES DEPENDING ON TRAINING |
| PHYSICAL TRAINING | 0535 | CONDUCTED LESS THAN 5 TIMES AT LDAC |
| BREAKFAST | 0630 | CADET LEADERSHIP DETERMINES PLT ORDER |
| FORMATION | 0745 | CADET LEADERSHIP DETERMINES EXACT TIME |
| TRANSPORTATION TO TRAINING | 0800 | ADJUSTED BY TIME/DISTANCE TO TRAINING SITE |
| TRAINING | 0830 | |
| LUNCH | 1200 | VARIES |
| TRAINING | 1230 | |
| DINNER | 1715 | VARIES |
| REGIMENTAL OPORD PROCESS | 1800 | OR 30 MINUTES AFTER DINNER MEAL (LEADERSHIP) |
| PREPARATION FOR NEXT DAY | 1800 | |
| LIGHTS OUT | 2200 | THIS TIME **NEVER** CHANGES |

**Leadership Insight 17:** Cadets who ride on the bus sit two to a seat and have their rucksacks in their laps during travel. Splitting the platoon leader and the platoon sergeant between two vehicles is a great technique for maintaining command and control (C2). Don't forget, the platoon sergeant is in charge of accountability, which is a major problem for most Cadets. During movement, the platoon leader should think ahead and be prepared for the next event (e.g. telling the PSG where to stage personnel and equipment, addressing the Cadre for specific instructions and risk mitigation).

**Leadership Insight 18:** Company level leadership evaluations begin when you receive the WARNO 2 days out. I suggest taking notes! Immediately after receiving the Regimental FRAGO, the clock begins to tick so Cadets must adhere to the one-third / two-thirds rule to afford subordinate units the time to develop their Operation Orders. (e.g. if you receive your FRAGO at 1900 and lights out is 2200, you have approximately 1 hour to finish planning and disseminate the company OPORD to the Platoon Leaders; most of this OPORD should be completed before receiving the FRAGO.)

**Leadership Insight 19:** Follow *"The Knuckle Dragger"* Approach. A common term in the Army is "Knuckle Dragger." Knuckle Dragger refers to individuals who don't have a large frontal lobe but can carry a hundred pound rucksack like there is no tomorrow. Knuckle Draggers do not have time for elaborate plans or multiple moving parts. They want to see pretty pictures that explain what is expected of them without a lot of thought. If you incorporate some key components in the OPORD process, you will follow the "Knuckle Dragger" Approach. The key components to receiving a good evaluation during the TLP process are: 1) Format – The use of the 5 paragraph OPORD format is a must so Cadets can issue a logical and natural flow of information to their subordinate leaders. The use of this format is the Army's way to communicate to lower echelons and it is very effective. 2) Graphics – Graphics assist the Cadet in disseminating information. Studies show that most individuals are visual learners; graphics are visual aids that allow Cadets to learn the order visually. At a minimum, use a map or draw a sketch to show the overall concept of the mission, write down the specified tasks and assign a squad to be responsible for that task, write out a timeline and construct a uniform and rucksack packing list. These graphics are best displayed on butcher block; an inherent task is to ask for butcher block if there is not one provided. Finally, don't forget to write these tidbits inside the Cadet TACSOP before leaving for LDAC. 3) One-third / two-thirds Rule – Following a time standard is a must or else Cadets at the lower levels will not be afforded the opportunity to develop a reasonable plan. Failure to follow the one-third / two-thirds rule will trickle down the chain and cause an emotional event for the squad leaders.

**Leadership Insight 20:** Some training events such as the land navigation training are a must pass event. The confidence training is a must attempt event. Failure to attempt an obstacle will result in a counseling statement for "Failure to Train."

**Leadership Insight 21:** During Garrison evaluations, eleven different Leadership Performance Indicators (LPI) are almost always seen. (Refer to Table 1.)

**Leadership Insight 22:** Most Cadets make the mistake of writing down the platoon mission statement instead of formulating a squad mission statement.

**Leadership Insight 23:** The majority of Cadets fail to plan for an EPW or CCP since they think the platoon CCP is the correct location because it is given to them; this is not the case. When producing the Squad OPORD, it is important to establish a CCP closer to the objective in case your squad receives casualties. After treating the casualties and return to the Objective Rally Point (ORP), they would then be evacuated to the platoon CCP.

**Leadership Insight 24:** Information gained from the enemy portion of paragraph I should enable leaders to formulate the Enemy Most Probable (EMP) Course of Action (COA) and the Enemy Most Dangerous (EMD) COA. Keep it straight and to the point!

**Leadership Insight 25:** The mission statement must include the 5 Ws: Who, What, When, Where and Why. Your mission statement is a combination of the information from the Platoon's Mission Stement and the Platoon's Task to Maneuver Units (this pertinent information is underlined).

*Who: 1st Sqd*

*What:* ***conducts*** *point ambush*

*When:* ***NLT*** *1500 15 May 2008.*

*Where:* ***IVO*** *ET 123456*

*Why:* ***IOT*** *prevent enemy forces from locating avenues of approach into BP Yankee.*

*1st SQD* ***conducts*** *point ambush* ***NLT*** *1500 15 May 2008* ***IVO*** *(in vicinity of) ET 123456 IOT (in order to) prevent the enemy from locating avenues of approach into BP YANKEE.*

**Leadership Insight 26:** Although the OPORD states cross the LD NLT H+45 and conduct ambush NLT H+80, you must convert this to actual time. You accomplish this by listening to the time the TAC gives you (e.g. "The time now is 13:40, what are your questions?") 13:40 becomes your base time for everything for this operation. Cross LD at 1425 = 13:40 + 45 min; Conduct Ambush at 15:00 =13:40 + 80 min.

**Leadership Insight 27:** Divide the maneuver paragraph into these 5-phases.

**Leadership Insight 28:** Keep the plan simple; don't develop an elaborate plan that is hard to understand. A simple plan executed violently is more successful than a complex plan that is never executed. Remember they acronym KISS (Keep It Simple Stupid).

**Leadership Insight 29:** Don't cross the LD early or late; use rehearsals from the time you complete the OPORD until it is time to cross the LD.

**Leadership Insight 30:** The squad column fire team wedge or the file formation is used the majority of the time during SSTX lanes.

**Leadership Insight 31:** Generally, if you have a good boot fit, you should have a slight amount of room in the front of the boots to wiggle your toes. Width-wise, your boots should feel a little snug (but not tight) and your heel should stay in place while walking and not slide back and forth (which causes blisters). To improve the fit of the boot, try loosening or tightening the laces at different parts of the boot until you get the perfect fit.

**Leadership Insight 32:** Change your socks at LDAC on a regular basis and use foot powder. Your feet will get wet at LDAC so changing your socks is a must and allows you to keep the second pair of boots dry once you arrive at the Assembly Area (AA) or Forward Operational Base (FOB).

**Leadership Insight 33:** Cadets must conduct foot marches to break in their boots and determine where blisters will form. At LDAC a Cadet walks approximately 75 km while conducting land navigation, foot marches, SSTX, movement to the FOB and PSTXs.

**Leadership Insight 34:** The best blister prevention is to get out and toughen up your feet by conducting at least two foot marches of at least five miles per month. Don't be a liability to your platoon at LDAC. Prepare yourself!

**Leadership Insight 35:** Use the hand sanitizer and wash your hands at every opportunity to avoid germs. Don't be afraid to go on sick call if you need sanitizer or medicine. Go see the doctor for spider, chiggers and mosquito bites before they become infected; use bug spray because the mosquitoes are HUGE!

**Leadership Insight 36:** Take the GRE/GMAT immediately following graduation. You will be more prepared to take it now than you will in three years. If you bomb it, you can take it again in six months. Nevertheless, I would not accept no for an answer because many universities and colleges are very patriotic and can be persuaded to accept you as a student. Ask for a waiver!

# Acronym List

| | |
|---|---|
| 1SG | First Sergeant |
| AA | Assembly Area |
| AAR | After Actions Review |
| ACE | Ammunition / Casualties / Equipment |
| ACS | Advanced Civil Schooling |
| ACU | Army Combat Uniform |
| ADS | Active Duty |
| ADSO | Active Duty Service Obligation |
| AKO | Army Knowledge Online |
| AMEDD | Army Medical Department |
| AO | Area of Operations |
| AOO | Actions of the Objective |
| APFT | Army Physical Fitness Test |
| AS | Assessing |
| AT | Anti-Tank |
| BD | Building |
| BRM | Basic Rifle Marksmanship |
| C2 | Command and Control |
| CB | Civilian on the Battlefield |
| CCIMS | Cadet Command Information |

| | |
|---|---|
| | Management System |
| CCP | Casualty Collection Point |
| CCR | Cadet Command Circulative |
| CENCO | Company Evaluation NCO |
| CEO | Company Evaluation Officer |
| CER | Cadet Evaluation Report |
| CIF | Central Issuing Facility |
| CN | Conceptual |
| CO | Communication |
| CO | Commander |
| COA | Course of Action |
| COC | Commandant of the Cadets |
| CP | Command Post |
| CTLT | Cadet Troop Leading Training |
| CTO | Company TAC Officer |
| CULP | Cultural Understanding and Language Program |
| CVG | Caquetian Volunteer Guard |
| CWST | Combat Water Survival Test |
| DA | Department of the Army |

| | |
|---|---|
| DE | Developing |
| DM | Decision Making |
| DMG | Distinguished Military Graduate |
| DX | Direct Exchange |
| ECP | Early Commissioning Program |
| EM | Emotional |
| EPW | Enemy Prisoner of War |
| EX | Executing |
| FLAR | Flat Ass Rule |
| FLEP | Funded Legal Education Program |
| FLRC | Field Leadership Reactionary Course |
| FM | Field Manuals |
| FOB | Forward Operating Base |
| FRAGO | Fragmentation Order |
| FTX | Field Training Exercise |
| GFRD | Guaranteed Reserve Forces Duty |
| GMAT | Graduate Management Admission Test |
| GOTWA | Going / Others / Time / Where / Actions on Enemy Contact |
| GPA | Grade Point Average |

| | |
|---|---|
| GPS | Global Positioning System |
| GRE | Graduate Records Examination |
| HRT | Human Resources Technician |
| IAW | In Accordance With |
| ID | Identification |
| IOT | In order To |
| IP | Inter-Personal |
| IPB | Intelligence Preparation of the Battlefield |
| IR | Information Requirements |
| ISO | In Support Of |
| ITT | Individual Tactical Training |
| IVO | In Vicinity Of |
| JPAS | Job Performance Summary Card |
| LBE | Load Bearing Equipment |
| LBE | Load Bearing Equipment |
| LD | Line of Departure |
| LDAC | Leadership Development and Assessment Course |
| LDP | Leadership Development Program |
| LE | Learning |

| | |
|---|---|
| LPI | Leadership Performance Indicators |
| LSD | Long Slow Distance |
| LOA | Limit of Advance |
| LT | Lieutenant |
| LTC | Leadership Training Course |
| MAJ | Major |
| MDCOA | Most Dangerous Course of Action |
| ME | Mental |
| MEDEVAC | Medical Evacuation |
| METL | Mission Essential Task List |
| METT-TC | Mission, Enemy, Terrain, Time, Troops Available, Civilian Considerations |
| MIA | Missing in Action |
| MO | Motivating |
| MPCOA | Most Probably Course of Action |
| MRE | Meals Ready to Eat |
| MS | Military Science |
| MSC | Medical Services Corps |
| NBC | Nuclear, Biological, Chemical |
| NCO | Non-Commissioned Officers |

| | |
|---|---|
| NLA | National Liberation Army |
| NLT | No Later Than |
| OAKOC | Observation & Fields of Fire, Avenues of Approach, Key Terrain, Obstacles and Movement, Cover & Concealment |
| OCOKA | Observation & Fields of Fire, Cover & Concealment, Obstacles and Movement, Key Terrain,Avenues of Approach |
| OCS | Officer Candidate School |
| OE | Operational Environment |
| OER | Officer Evaluation Report |
| OML | Order of Merit List |
| OMS | Order of Merit Score |
| ONA | Overall Net Assessments |
| OOM | Order of Movement |
| OPFOR | Opposing Forces |
| OPORD | Operations Orders |
| OPS | On Point Saliency |
| ORP | Objective Rally Point |
| PB | Patrol Base |
| PCC | Pre-Combat Checks |

| | |
|---|---|
| PCI | Pre-Combat Inspections |
| PDF | Primary Direction of Fire |
| PERSTAT | Personnel Status |
| PH | Physical |
| PIR | Priority Intelligence Requirements |
| PL | Platoon Leader |
| PL | Planning |
| PMI | Pre Marksmanship Instruction |
| PMS | Professor of Military Science |
| PSG | Platoon Sergeant |
| PSTX | Patrolling Situational Training Exercise |
| PT | Physical Training |
| PTLT | Platoon Tactical Lieutenant |
| PTNCO | Platoon Tactical Non-Commissioned Officer |
| PTO | Platoon Tactical Officer |
| PX | Post Exchange |
| REO | Regimental Executive Officer |
| RFI | Request For Information |
| ROE | Rules of Engagement |

| | |
|---|---|
| ROTC | Reserve Officer Training Corps |
| RP | Release Point |
| RSGM | Regimental Sergeant Major |
| RTO | Regimental TAC Officer |
| SAL | Student / Athlete / Scholar |
| SAW | Squad Automatic Weapon |
| SEA-TAC | Seattle-Tacoma |
| SF | Special Forces |
| SFC | Sergeant First Class |
| SFG | Special Forces Group |
| SFODA | Special Forces Operational Detachment – Alpha |
| SGM | Sergeant Major |
| SITREP | Situation Report |
| SL | Squad Leader |
| SOP | Standard Operating Procedures |
| SP | Start Point |
| SSTX | Squad Situational Training Exercise |
| STAR | Situation / Task / Action / Result |
| SUT | Small Unit Tactic |

| | |
|---|---|
| TA | Tactical |
| TAC | Train Advise Counsel |
| TACSOP | Tactical Standard Operating Procedures |
| TCM | Take Charge Minute |
| TCP | Traffic Control Point |
| TE | Technical |
| THS | Tiny Heart Syndrome |
| TIV | The Inner Villain |
| TLP | Troop Leading Procedures |
| TRP | Target Reference Point |
| TTP | Tactics / Techniques / Procedures |
| UCF | University of Central Florida |
| USD | United States Dollars |
| USMA | United States Military Academy |
| VECP | Vehicle Check Points |
| WARNO | Warning Order |
| WO | Warning Order |
| WP | Water Points |
| XO | Executive Officer |
| ZOS | Zone of Separation |

# References

## ARTICLES

Hewitt, Brian., (June 12, 2004.) *Boot Fitting Guide*
*www.greatoutdoors.com/published/boot-fitting-guide*

## CIRCULARS, PAMPHLETS, AND REGULATIONS

| | |
|---|---|
| CC Circ 145-05 | *Leader Development and Assessment Course (LDAC)/Warrior Forge (WF) Preparation*<br>15 December 2008 |
| CC Circ 601-10-1 | *ROTC Accessions Fiscal Year* (10)<br>2 March 2009 |
| CC Reg 145-1 | *Reserve Officers' Training Corps Army ROTC Incentives Policy*<br>7 May 2008 |
| CC Pam 145-1 | *ROTC Incentives Procedures*<br>14 March 2008 |
| CC Pam 145-3-2 | *Cultural Understanding and Language Proficiency*<br>14 August 2009 |

| | |
|---|---|
| CC Pay 145-4 | *Enrollment, Retention and Disenrollment Criteria Policy and Procedures*<br>21 August 2008 |
| CC Reg 145-3 | *Reserve Officers Training Corps Pre-commissioning Training and Leadership Development*<br>20 September 2006 |
| CC Pam 145-9 | *ROTC Cadet Professional Development Training*<br>10 March 2003 |
| CC Reg 145-9 | *Reserve Officers' Training Corps Accessioning and Commissioning*<br>3 March 2008 |

## FIELD MANUALS

| | |
|---|---|
| FM 3-21.8 | *The Infantry Rifle Platoon and Squad*<br>28 March 2007 |
| FM 3-21.10 | *The Infantry Rifle Company*<br>27 July 2006 |
| FM 3-21.75 | *The Warrior Ethos and Soldier Combat Skills*<br>28 January 2008 |
| FM 3-21.5 | *Drill and Ceremonies*<br>07 July 2003 |

| | |
|---|---|
| FM 3-22.9 | *Rifle Marksmanship, M16/M4 Series*<br>12 August 2008 |
| FM 3-25.26 | *Map Reading and Land Navigation*<br>18 January 2005 |
| FM 6-22 | *Army Leadership*<br>12 October 2006. |
| FM 7-8 | *Company and Platoon Infantry Manual*<br>22 April 1992 |
| FM 21-18 | *Foot Marches*<br>01 June 1990 |
| FM 21-20 | *Physical Fitness Training*<br>30 September 1992 |
| STP 21-1-SMCT | *Soldier Manual of Common Tasks*<br>18 June 2009 |

## HANDBOOKS

| | |
|---|---|
| N/A | *Leadership Development Program Handbook*<br>25 August 2009 |
| SH 21-76 | *Ranger Hand Book*<br>02 July 2006 |

## FORMS

| | |
|---|---|
| CC Form 156-2-R | *Self Assessment Card Report*<br>July 2009 |
| CC Form 156-4A-R | *Leadership Assessment Report*<br>July 2009 |
| CC Form156-18A-R | *Job Performance Summary Card-Cadet*<br>July 2009 |
| CC Form 156-18B-R | *Job Performance Summary Card-Cadre*<br>July 2009 |
| CC Form 156-100-R | *Leadership Performance Indicator Card*<br>July 2009 |